Human-Centered Artificial Intelligence for Industrial Safety

This book examines the transformative role of artificial intelligence (AI) in industrial safety, focusing on the shift from traditional reactive systems to proactive, predictive safety architectures in smart factories. It explores how AI technologies, including machine learning, computer vision, biometrics, wearables, augmented reality, and digital twins, are revolutionizing occupational health and safety across industries such as manufacturing, logistics, energy, aviation, and construction. This book highlights the integration of AI-enabled safety systems, addressing real-time risk monitoring, predictive maintenance, employee well-being, and accident prevention. It also delves into the ethical, legal, and regulatory implications of AI in safety-critical systems, including GDPR compliance, algorithm transparency, and ISO 45001 standards. Case studies from global companies like Siemens, Boeing, Samsung, and INVISTA provide practical insights into implementation, challenges, and outcomes.

Knowledge-based Engineering for Innovation

Nilanjan Dey, Professor with the Techno International New Town, Kolkata and visiting fellow of the University of Reading, UK

The series presents invaluable insights into knowledge-based engineering for innovation by investigating the role of emerging technologies like machine learning and artificial intelligence in enhancing knowledge management, data-driven research, knowledge-intensive applications, knowledge transfer, and related areas. By addressing the challenges and opportunities in this domain, the series aims to bridge theoretical and practical aspects through transdisciplinary approaches, highlighting the critical role of knowledge-driven research in today's academic and industrial landscapes. The series promotes the formation of knowledge-driven communities, particularly in the Information Technology sector, while also bridging diverse fields such as healthcare, education, social sciences, business, and environmental studies to this platform for examining strategies, tools, and best practices that support collaborative and impactful research.

Application of Computational Techniques in Power Systems for Power Quality Improvement
Jagatheesan Kaliannan, Boopathi Dhanasekaran, Sourav Samanta, Anand Baskar, Kanendra Naidu

Knowledge-based Engineering for Innovation
Ketan Sanjay Desale and Soumi Majumder

Innovation and Knowledge Management in Manufacturing Industry: Case Studies and Applications
Ayan Dutta and Bitan Misra

Human-Centred Artificial Intelligence for Industrial Safety
Joanna Rosak-Szyrocka, Radosław Wolniak, and Michalene Eva Grebski

www.routledge.com/Knowledge-based-Engineering-for-Innovation/book-series/KBEI

Human-Centered Artificial Intelligence for Industrial Safety

Joanna Rosak-Szyrocka
Radosław Wolniak
Michalene Eva Grebski

CRC Press
Taylor & Francis Group
Boca Raton London New York

CRC Press is an imprint of the
Taylor & Francis Group, an informa business
A CHAPMAN & HALL BOOK

Designed cover image: Shutterstock (Image ID: 2680956053)

First edition published 2027
by CRC Press
2385 NW Executive Center Drive, Suite 320, Boca Raton FL 33431

and by CRC Press
4 Park Square, Milton Park, Abingdon, Oxon, OX14 4RN

CRC Press is an imprint of Taylor & Francis Group, LLC

ISBN: 978-1-041-26952-6 (hbk)
ISBN: 978-1-041-27407-0 (pbk)
ISBN: 978-1-003-75421-3 (ebk)

DOI: 10.1201/9781003754213

Typeset in Times
by SPi Technologies India Pvt Ltd (Straive)

Contents

About the Authors

Joanna Rosak-Szyrocka is an assistant professor in the Faculty of Management at Częstochowa University of Technology. Her research interests focus on quality management and process improvement, increasingly linked to digital transformation in Industry 4.0 and Quality 4.0 concepts, as well as sustainability-oriented management topics such as ESG and responsible innovation. In her scientific output, she addresses organizational and technological determinants of efficiency and quality in manufacturing and service settings, often emphasizing how data-driven tools (e.g., digital quality systems, analytics, and AI-related solutions) can support continuous improvement and sustainable performance. She has been listed in Stanford University's ranking of the World's Top 2% Scientists.

Radosław Wolniak is a full professor at the Department of Economics and Informatics at the Silesian University of Technology (Poland). For over two decades, he has been dealing with the issues of quality management, Industry 4.0, and Smart City. He has been listed in Stanford University's ranking of the World's Top 2% Scientists.

Michalene Eva Grebski is an instructor of psychology at Colorado Mesa University. She teaches Psychology and Human Growth and Development courses. Michalene Eva Grebski has a multicultural educational background and believes in a holistic, multidisciplinary approach to education. In her classes, she promotes a democratized, risk-taking culture as well as lateral thinking.

Preface

The fast advancement of artificial intelligence, cyber-physical systems, and intelligent perception technologies is rapidly evolving current workplaces to be some high-coherent, transparent, and continuous learning structures. In this context, occupational safety becomes an output of companies' maturity based on AI focused on humans, the principles of Industry 5.0, and ESG and SD values. With this understanding, safety goes far beyond managing traditional risks and includes stable practices, predictive capabilities, operational fortitude, and the judicious use of data and automation.

In response to these challenges, the authors of this reference book have developed an overarching conceptual model called the Three-Ring Model – a vision that unites three critical dimensions of contemporary safety systems: predictive intelligence, human-augmentation technologies, and ethical-regulatory foundational frameworks guiding system performance. Such a model is useful to examine the design of work environments where technologies improve action coherence, add to decision transparency, build resilience, and also involve workers with long-term well-being, health, and engagement. It emphasizes the significance of responsibility, the integrity of process, and human–technology interaction design.

It is suitable for use by a wide range of parties who are responsible for achieving high-quality company functioning. It provides a rational tool for anyone designing data-based safety systems and supports engineers and experts in the fields of ergonomics or work environment development as well as teams involved in technological supervision, regulation compliance, and ethical introduction of new technology. From an organizational development standpoint, this model offers a framework to assess process alignment and address gaps in operational capacity; additionally, it helps in supporting informed decisions related to responsible and sustainable technology adoption. This is applicable at both strategic and operational levels – from defining policies and governing systems to the deployment of concrete AI solutions that make work safer and more reliable.

This book develops a perspective that safety is an organizational capacity based on the interplay of people, technologies, and decision structures. Such capacities contribute to an organization's long-term ability to create safe, healthy, and development-friendly working environments that improve resilience, predictability, and quality of work in the context of rapid technological change. This book is suitable for both researchers and professionals who are interested in developing mature, transparent, and human-centred safety systems that meet the challenges of Industry 5.0.

Introduction

As early as the beginning of modern AI, **Andrew Ng has said that 'AI is the new electricity'**, an infrastructural technology that not only powers but recasts societies and organizations from their base. When set against the background of this reference book, the metaphor is equally striking: just as electricity formed the basis for twentieth-century industrialization, artificial intelligence now stands out as one of the vital factors underwriting occupational safety in cyber-physical systems and Industry 5.0.

The revolution of industry by AI digitalization, cyber-physical systems, and sustainable production has led to a new concept of occupational safety. This is a paradigm shift from the traditional kind of safety based on control, reactive procedures, and hierarchical supervision to one that encompasses predictive, adaptive, and deeply human-centred ways of being. This book analyses such aspects from a multidimensional standpoint, laying down its theoretical, technological, and ethical-regulatory framework.

While the previous work provides detailed reflections of single concepts in AI safety, such as predictive analytics, ergonomic technologies, or ethical and regulatory frameworks, they seldom pull these dimensions together into a holistic conceptual model. Current approaches continue to be fragmented and focus on particular aspects of technologies or organizations; they are therefore incapable of reflecting the sociotechnical complexity of safety in modern industrial environments. This book seeks to fill this gap by advocating an integrative, human-centric view that conjoins predictive intelligence, human augmentation, and ethical governance from the beginning and is consequently based upon empirical evidence as well as real-world industrial applications.

The aim of this book is therefore to build a multidimensional, evidence-based picture around AI and industrial safety that reflects the potential afforded by technologies and the ethical, social, and organizational limitations of AI use. The Three-Ring Model developed by the authors of this book is a unique contribution to the existing literature, both in theoretical and practical terms – not only do they include examples of applications but also methods for analysis and design. In this sense, the reference book serves as both a theoretical backdrop for academics and as an applied model for practitioners who are seeking to develop AI in work organizations.

This book contributes to quality and management science by developing an integrated, evidence-based approach that sharpens occupational safety as a function of digital transformation. Through the orchestration of predictive analytics, human augmentation, and ethical governance, it illustrates how businesses can create computing systems that are exceedingly resilient, transparent, and performant. It challenges the reader to create new management paradigms that respect quality and safety as strategic capabilities in Industry 5.0.

This book tackles a number of core research questions that are key to comprehending human-centred AI for industrial safety:

- What are the long-term effects of advanced predictive AI technologies on early hazard prediction and risk mitigation in the industrial context?
- What are the ways in which augmentative technologies increase the physical and cognitive capacities of people and reform human experience in factory labour?
- What ethical, regulatory, and governance mechanisms are required to ensure AI systems retain transparency, accountability, and social licence in industrial settings?
- How might these three areas, predictive intelligence, human augmentation, and ethical governance, be combined into a universal model to enable industrial safety for humans?

This book consists of 10 chapters. In Chapter 1, the historical and conceptual development of safety: from Safety I to Safety 5.0, where human–AI cooperation is essential in HITL and HOTL, is considered. The authors reveal the change of a worker: from being 'the weakest link', he becomes a coauthor of decisions. The chapter is completed with two case studies, Protex AI and Surveily, showing how visual analytics supported by AI turns safety into intelligence; organizations are moving from traditional reactive systems to learning ones that can prevent accidents at the earliest stage. Chapter 2 presents the role of machine learning, supervised and unsupervised algorithms, reinforcement learning, and semi-supervised approaches in anomaly detection. At the core of the debate stand digital twins, supported by means of an Edge–Fog–Cloud architecture, providing the platform for multi-layered and adaptable safety system opportunities. The Siemens MindSphere case study is presented in the chapter and demonstrates how the predictive models can be merged with digital twins to minimize failures, operational mistakes, and downtime. Chapter 3 discusses the evolution of classical sensor data into cognitive, self-learning sensor networks (ISNs), environmental sensors, biomechanical sensors, and machine-related sensors. Through edge analytics and superior data fusion, such systems are 'able to "understand" the context' and real-time risk prognosis. Many industrial applications demonstrate how smart perception enables predictive maintenance and autonomous hazard detection, as well as more resilient organizations. Chapter 4 is dedicated to wearables, including smart insoles, sensor helmets, biometric vests, and RTLS systems for fatigue assessment, fall detection, overload monitoring, or posture analysis. The chapter discusses ethical and privacy concerns, in addition to the issue of 'alert fatigue'. The experience from INVISTA unveils that the deployment of RTLS and wearables reduces time to evacuate, enhances emergency response, and decreases the number of incidents. Chapter 5 considers the use of AR-enhanced interfaces to present instructions, warnings, and technical information overlaid onto the worker's working environment. AR minimizes mistakes, reduces overall surgery time, improves procedural adherence, and enhances spatial understanding. The chapter presents some examples in industry, namely, Boeing wire-harness assembly, DHL Pick-by-Vision, and the DAQRI Smart Helmet, for which quality increase, time of operation decrease, and risk reduction are reported. Chapter 6 presents a major development in ergonomics: AI-based exoskeletons that track real-time biomechanics (IMU, EMG, and force sensors), predict fatigue, and modulate the level of support accordingly. Included in the conversation are predictive models, multimodal sensing, and biomechanical digital twins. The chapter describes the Boeing–EksoVest example, where ergonomic data led to the development of advanced exoskeletons that adapted support levels so as to lift more than 30% of loads. In Chapter 7, we discuss the application of deep neural networks for PPE detection, unsafe behaviour analysis, and building predictive risk models based on visual data. The authors consider the complete CV pipeline: data collection, annotation, object detection, multimodal fusion, and temporal modelling. Two specific case studies will be presented: Protex AI + Marks & Spencer and Samsung Smart Factory, which, just to mention two of them, are illustrations of how the use of AI vision helps reduce occurrences and foster ergonomics as well as decision-making. In Chapter 8, the author also presents real-time localization systems and trajectory prediction algorithms for human–robot and robot–machine collision-free movement. Here is where the idea of dynamic risk maps comes into play, with hazard zones constantly updated. RTLS not only serves as a security tool but also optimizes flow, ergonomics, spatial management, and digital twin analysis. Chapter 9 also addresses the fundamentals of algorithmic accountability, explainability (XAI), data documentation, GDPR compliance, ISO 45001, and the EU AI Act. It examines how institutions can build an AI system, principles of proportionality and transparency, data protection, and institutional responsibility. The chapter also explains how XAI can facilitate audits, minimize bias, and reinforce employee confidence. Chapter 10 emphasizes the empathic, self-adjusting ESG-based safety of tomorrow. It explores applications for emotional AI, adaptive workstations, psychophysiological monitoring, and human-centric robotics. Good examples like Nissan Intelligent Factory, HP–Virgin Pulse, and the Siemens DEGREE Framework show that safety is increasingly seen as a responsible organization's strategic asset.

The authors of this book put forward a novel Three-Ring Model (predictive intelligence, human augmentation, and ethical governance) as an integrated conceptual construct to represent HCI-AI for Industrial Safety (Figure 0.1).

The model is one of the first constitutions that systematically provides a framework for the coordination between the predictive AI infrastructures and human augmentation technologies, together with ethical-organizational governance within a single integrated human-centred safety system. This model structures current AI-enabled safety approaches and encompasses the experimental results and theoretical understanding discussed in this book's chapters.

It is a dimensional analytical framework to conceptualize the way in which AI can be supportive of industrial safety as long as it has been conceived and used in a manifest human-centred approach.

The model posits that effective, responsible, and socially legitimate uses of AI at work will be the result of the interdependent development of three linked domains: predictive analytic systems for early hazard discovery and appraisal; technologies for enhancing human physical and cognitive capabilities; and institutional governance frameworks that underpin transparency, trustworthiness, and

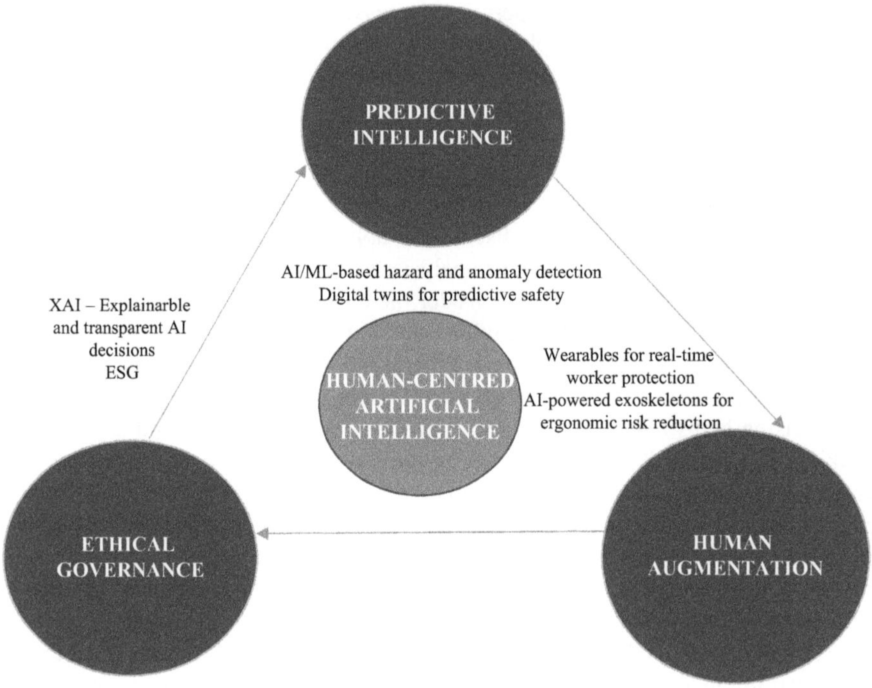

FIGURE 0.1 Three-Ring Model (predictive intelligence, human augmentation, and ethical governance) for industrial safety.

(Own study.)

ethical integrity. The human actor, positioned at the centre of this system, is the focus for well-being, autonomy, and decision-making capacity that define how AI shall be designed and deployed.

The innermost ring, predictive intelligence, is the technology heart of the model and includes predictive analytic tools such as machine learning, anomaly detection, sensor-fusion pipelines, and digital twins. These answers not only allow for the detection of risk factors or threats in situ but also serve to simulate risks and make and establish preventive operating suggestions. The analysis reveals that, in this book, a high level of machine perception technology is considered to enhance the flexibility and accuracy of hazard detection at workplaces, thereby enhancing the effectiveness of operational safety management. It is hence the predictive intelligence that becomes the data-driven backbone of contemporary safety systems by improving responsiveness and analytical extensiveness.

The second ring is called human augmentation, which includes technologies that strengthen human abilities and parallel cognitive and biomechanical loads. This space encompasses wearables (monitoring physiology and behaviour), AR systems where real-time perceptual augmentation takes place, and algorithmically controlled exoskeletons that can tune support in response to the demands of the task. The analyses in this book show that assistive technologies are essential for minimizing risk of injury, facilitating optimal ergonomic conditions, and supporting high levels of human functional performance while working under challenging operational conditions. Human augmentation therefore recasts the human–machine relationship from one that is based on overseeing and command to mutual enablement, whereby AI infrastructure empowers human safety, awareness, and productivity.

The third ring, ethical governance, consists of the normative and institutional safeguards to ensure responsible deployment of AI in workplace scenarios. Model transparency (XAI), algorithmic accountability, privacy protection, system auditability, and alignment with ESG frameworks are key to this dimension and increasingly define how organizations think about safety and sustainable development. This book's analyses underscore how deploying AI without strong ethical underpinnings results in lower employee trust and greater organizational resistance, as well as a greater potential for unintended consequences. So, governance works more as a stabilizing layer, helping to make sure that predictive and augmentative technology behaves in proportional, transparent ways according to organizational values and employee rights.

This three-ring synergy constitutes a comprehensive structure to analyse, design, and implement AI-based safety systems. Predictive intelligence gives organizations the potential for advanced hazard notification and decision support; human augmentation enhances capabilities of humans to diminish risk exposure; and ethical governance supports responsible, auditable, transparent use. Their combination results in a unified model, with AI as an infrastructure that guarantees human safety, well-being, and autonomy across industrial places.

We expect that this book will provide readers with an enticing and intellectually challenging research journey and lead to enhanced insights on how AI can shape the future of occupational safety.

Joanna Rosak-Szyrocka
Radosław Wolniak
Michalene Eva Grebski

Part I

Foundations of AI-Driven Safety in Smart Factory

1 Workplace Safety in the AI Era

1.1 THE EVOLUTION OF SAFETY SYSTEMS: FROM REACTIVE TO PREDICTIVE

Each year, about 2 million people die from work-related causes (WHO & ILO, 2021). In 2023, there were 2.82 million non-fatal accidents at work that resulted in at least 4 calendar days of absence from work and 3,298 fatal accidents at work in the EU (Eurostat, 2024). The United Nations has announced 17 Sustainable Development Goals (SDGs) and 169 objectives (Boto-Álvarez & García-Fernández, 2020) in its Agenda 2030 for Sustainable Development (Szopik-Depczyńska et al., 2018), some of which have to do with workplace safety and health: the third SDG is to 'ensure healthy lives and promote well-being for all at all ages'. The eighth SDG is to 'promote sustained, inclusive and sustainable economic growth, full and productive employment, and decent work for all'. The sixteenth SDG is to 'promote peaceful and inclusive societies for sustainable development, build effective, accountable, and inclusive institutions at all levels, and provide access to justice for all'. When implementing occupational safety and health (OSH) regulations and incentive programmes at various levels and geographic scales throughout this period, these objectives, as well as the public health goals, must be taken into account (Ávila-Gutiérrez et al., 2022). These findings emphasize the increasing relevance of systemic and preventive approaches to occupational safety, in line with the principles of SDGs decent work and well-being.

Safety evolution in the age of smart factories mirrors a deep-seated shift in thinking about how organizations view the way humans, technology, and data are used to keep production processes online and running smoothly (Maia et al., 2022; Rosak-Szyrocka et al., 2025; Rosak-Szyrocka & Wolniak, 2025b). Cyber-physical systems are becoming more complex, as automation advances and artificial intelligence (AI) enters the setting, such that classically modelled safety through control structures and hierarchy, as well as post-hoc response to incidents, is ineffective for these new interconnected industrial ecologies (Hollnagel et al., 2017; Leveson, 2016).

Modern understandings of safety highlight the dynamic and emergent nature through which safety emanates from interactions among humans, algorithms, and technical infrastructures. From a contemporary point of view, safety is no longer described in terms of being static but as an active and continuous learning and adaptive procedure dynamically co-produced between humans and technologies within conditions of operational uncertainty (Ávila-Gutiérrez et al., 2022; Sarvari et al., 2024).

It has been noted in recent literature that such evolution compels the need to redefine the basic principles of the safety function. There are three primary theoretical lenses of safety: Safety I, Safety II, and Safety III – that collectively explain the progression in which safety has been conceptualized within socio-technical systems

DOI: 10.1201/9781003754213-2

(Hollnagel, 2018; Leveson, 2016). These paradigms represent a conceptual transition from reactive, control-based perspectives to proactive, system-focused, and ultimately humanistic and ethical viewpoints where safety is established as an organizational value that is shared by all and considered as part of sustainable development (Afonso-Fernandes et al., 2026; Liu et al., 2024). Table 1.1 presents the evolution of safety paradigms from Safety I to Safety III.

TABLE 1.1
Evolution of Safety Paradigms from Safety I to Safety III

Paradigm	Definition and Purpose	Human Role	Organizational Approach
Safety I	Safety means the absence of accidents, errors, and mistakes. It rests on the premise that each unwanted occurrence has a cause, and if action is taken to remove the cause, then the undesired event will not occur. The primary objective is to control, obey, and avoid human mistakes.	The human is viewed as the vulnerable link in a socio-technical system, subject to error, requiring monitoring and discipline through surveillance and standardization.	A compliance-oriented, reactive, and rule-dominated safety culture that focuses on hierarchy, discipline, and accountability through blame. It is not by accident that organizations depend on incident reporting, root cause analyses, and regulatory enforcement for safety.
Safety II	Safety refers to the system's ability to operate successfully in a variable, uncertain environment. It is specifically interested in learning from everyday performance ('work-as-done') and enhancing resilience and adaptability.	The human is thought of as an adaptive, resilient resource that can effectively react to surprises and variety. Human expertise, situational awareness, and creativity for system robustness.	A safety culture that is proactive and learning based, founded on trust, open communication, and continual improvement. Institutions promote feedback, embracing successes and collective safety responsibility at all levels.
Safety III	Safety is understood as a value co-creation process that incorporates moral, social, and environmental aspects in human–AI–organization systems. It is focused on transparency, algorithmic accountability, and human welfare.	Humans and AI are co-decision makers in human-in-the-loop (HITL) and human-on-the-loop settings. The human provides ethical supervision and contextual understanding for AI-based decisions.	Ethically and people-centred, trust-based safety culture that encourages cooperation and transparency of data, with the aid of a common moral duty. Safety becomes a core company value just as essential as sustainability, and the well-being of employees once again takes on top priority.

Source: Own study.

Safety I became predominant in the latter half of the nineteenth century and throughout most of the twentieth century, when increasing mechanization and early automation required managing accidents and standardizing working practices. Safety was conceived as the non-occurrence of accidents, errors, and incidents, where the goal is to gain an accident-free performance by analysing and removing their causes (Hollnagel, 2018). Safety I infrastructure is based on a linear model of causality, premised upon the conviction that each negative event has one causal factor or condition assignable as root cause. Key techniques involved were root cause analysis (RCA), corrective action, and very firm compliance controls. In this scheme, humans were the weakest link in the chain: something liable to error and had to be monitored and controlled. Although such philosophy played a key role in shaping classical safety culture, which was discipline based and compliance driven, its limitations in the complexity of socio-technical systems, where interactions among humans, technology, and organizations became less predictable (Hollnagel et al., 2017; Leveson, 2016; Sarvari et al., 2024), were apparent.

By the early twenty-first century, a different understanding of safety had developed – which came to be known as Safety II. This is epitomized by what has been described elsewhere as Resilience Engineering, a concept conceived by Erik Hollnagel (Hollnagel, 2018) in reaction to the limitations of the reactive paradigm that sees a system's safety in its capacity to operate successfully under changing and unpredictable circumstances rather than merely avoiding incidents (Ávila-Gutiérrez et al., 2022; Hollnagel, 2018; Hollnagel et al., 2017).

Rather than searching for the analysis of failures, Safety II seeks to learn from what happens when things go right. This is a paradigm shift from a reactive to a resilience-based approach to safety, in which variability and adaptability are seen as resources. We no longer consider humans as threats but as adaptive units that help the systems to remain stable and persistent despite disturbances. This model encourages us to manage risk proactively, learn continuously, and communicate openly at all levels of the organization. New research (Ávila-Gutiérrez et al., 2022; Liu et al., 2024; Rodriguez et al., 2025) has shown that Safety II now underpins current safety approaches in complex systems: enabling organizations to foresee and take up operational uncertainty and vulnerabilities.

The latest and emerging new theory, Safety III, adds a whole dimension to safety thinking towards the incorporation of ethical, social, and ecological factors in the argumentation of safety. Contrary to Safety II, which is based on human adaptability and system resilience, Safety III directly integrates moral reasoning, algorithmic transparency, and the human–AI co-responsibility for safe decision-making. It brings a human-oriented view where safety is not considered as static or end-state, but as value co-creation among humans, technology, and organization's interactions (Aven, 2022; Ávila-Gutiérrez et al., 2022; Leveson, 2016).

The latest contributions (Samost-Williams & Brook, 2025; Sarvari et al., 2024) indicate that this approach could establish a framework conducive to the ethics as well as socially responsible reform of organizations in the AI era (Rosak-Szyrocka et al., 2021; Rosak-Szyrocka & Wolniak, 2025a; Rosak-Szyrocka & Wolniak, 2025b; Tripathi & Rosak-Szyrocka, 2024).

Safety in the workplace today has become one of the two essential axes and a prerequisite for sustainable development of organizations, as well as stability over time of an industrial establishment (Lemos et al., 2024; Ramírez-Peña et al., 2025; Rodriguez et al., 2025). It is even more important now in the era of digitalization, automation, and exponential growth in AI as its scope extends far beyond regulatory requirements and post-incident management (Chia et al., 2019; Scandelai, 2025). Contemporary safety is understood as an evolving system that relies on data-informed procedures, risk advanced warning, and human–technology interplay (Ávila-Gutiérrez et al., 2022; Bartulović, 2021).

Safety system development has traditionally advanced at the same pace as industrial revolutions (Abouhawwash et al., 2024; Rosak-Szyrocka, 2025a). Each revolution changed not only production systems but also the way in which businesses thought about the relationship among man, machine, and workplace (Rosak-Szyrocka et al., 2021; Rosak-Szyrocka et al., 2023b). In this context, safety concepts have been more and more addressed by scholars, which reflects the development of paradigm shifts in risk management, as well as in organizational learning (Chia et al., 2019; Min et al., 2019; Musarat et al., 2023; Rosak-Szyrocka & Wolniak, 2025a). Table 1.2 presents the evolution of safety systems from Safety 1.0 to Safety 5.0.

Safety 1.0 was characteristic of the First Industrial Revolution and was built around reacting to things – taking action after something had already happened. The overriding discourse was technological, with safety framed as machine guarding and accident prevention rather than systemic risk control. There were few preventive practices, promoted procedures, or an integrated safety culture (ICAO, 2013). In safety systems end at the line above were mainly based on End of Line inspection and corrective actions, with a resultant reactive and static system.

During the Second Industrial Revolution, when mass production was growing, Safety 2.0 appeared, introducing the first preventive measures and standardization of components. Regular inspections, worker education, and the formation of industrial safety regulations become cornerstones of business regulation. The focus on safety turned from individuals to groups, and thus, it was the beginning of a systematic approach to risk prevention (Fink, 2010). This stage marked a clear shift from post-event correction to preventive action.

This change was pushed through in the second half of the twentieth century due to automation and process-based management. This time frame saw the emergence of Safety 3.0, which accentuated human and organizational factors and included mainstreaming the idea of a Safety Management System (SMS; Mohy et al., 2024). Here, a more proactive approach was initiated – identifying potential risks early and addressing root causes prior to accidents (Liu et al., 2020).

The so-called Fourth Industrial Revolution with its digital technologies led to Safety 4.0. It is built upon streamlining actionable data-led analytics colliding with the marriage of traditional safety management mixed with Industry 4.0, which includes the Internet of Things (IoT), artificial intelligence (AI), machine learning (ML), and big data. Control systems became safety systems and beyond as they gained intelligence to monitor and recognize patterns and take autonomous preventive action.

TABLE 1.2
The Evolution of Safety Systems from Safety 1.0 to Safety 5.0

Safety System	Concept	Description	Tools and Technologies
Safety 1.0 – reactive safety	End-of-line control/ accident response.	Safety and lack of accidents or injuries. Based on post-incident analysis and simple hazard reduction. Human error is identified as the primary cause of accidents. No preventive measures, little regulation, and little system standardization.	Manual inspection, mechanical guards, and visual checks.
Safety 2.0 – preventive safety	Prevention through standardization and regulation.	Safety in terms of no accidents or injuries. More focused on remediation and primary hazard control in the aftermath of an incident. Focused on prevention by standardization, inspection, and early regulation.	Checklists, safety audits, statistical control, safety training programmes, and early safety codes.
Safety 3.0 – systemic safety	Safety Management Systems (SMS) and human factors.	Introduction of a systemic and proactive approach. Combined human, organizational, and technology factors. Concentrate on the lessons learned from near-misses and create a culture of safety.	Safety Management Systems (SMS), Supervisory Control and Data Acquisition (SCADA), Root Cause Analysis (RCA), Behavioural Safety Programmes, and Key Performance Indicator monitoring.
Safety 4.0 – predictive and intelligent safety	AI-driven, data-informed, real-time safety.	Combining digital competence and technologies (IoT, AI, and Big Data) with safety management. Health and safety become data-informed, proactive – even pre-emptive of risk prior to incidents happening. Systems are learning in a self-learning manner and are acting in real time.	IoT sensors, AI-based predictive analytics, digital twins, machine learning, wearables, MES, and SCADA 2.0.
Safety 5.0 – human-centric and sustainable safety	Collaborative human–AI and ethical safety systems.	The latest ethical and technology and human well-being universe. Safety as a value co-creation process among humans, AI, and organizations. Develop trust in data, transparency, and sustainability.	Human-in-the-loop and human-on-the-loop systems, ethical AI, neuroergonomics, AR/VR training, and digital safety culture platforms.

Source: Own study.

The current stage, also known as Safety 5.0, can be regarded as an evolution of the predictive paradigm and is in line with the general concept of Industry 5.0 that connects advanced technologies with values related to a human-centric approach, data ethics, and sustainability (Kumar et al., 2024; Rosak-Szyrocka & Knop, 2024; Rosak-Szyrocka et al., 2023a). The goal is no longer about minimizing accidents but also to establish workplaces, where well-being, trust, and joint responsibility are present (Bennouna et al., 2025; Berx et al., 2025; Rosak-Szyrocka et al., 2025; Rosak-Szyrocka & Knop, 2024; Rosak-Szyrocka & Tiwari, 2023). The role of humans is not anymore as the weakest link in the system, but they are an active partner, co-creator of safety decisions backed by AI systems (McCaffrey et al., 2025; Rosak-Szyrocka & Wolniak, 2025a). Here, safety gets an ethical and social dimension and becomes one of the pillars of sustainable technology advancement.

The shift to Safety 5.0 has also led to the development of a new dimension of safety culture – digital safety culture based on trust, data transparency, and organizational learning. It is a culture where increasingly safety decisions are underpinned by predictive analytics, and risk management becomes collaborative among humans, technology, and organizations (Bartulović, 2021). This transformation represents the redefinition of safety management as an adaptive learning system integrated within digital corporate ecosystems.

Under the banner of Safety 5.0, there is a growing focus on sustainable safety, a lasting and comprehensive model that encompasses human protection, environmental preservation, and organizational resilience. Contrary to the conventional models, centred mainly on accident decrease, sustainable safety encompasses the reconciliation of safety management with the broader aims of sustainability – health, well-being, and technological responsibility (Ávila-Gutiérrez et al., 2022; Gamberini & Pluchino, 2024; Pasman & Behie, 2024). From this angle, safety becomes a permanent value within the organization and a fundamental support for continuous growth, in which technology decisions are made openly, considering their social, ethical, and environmental implications (Rame et al., 2024; Wang et al., 2023; Yitmen et al., 2024).

1.2 HUMAN-CENTERED SAFETY IN INDUSTRY 4.0 AND 5.0

As the concept of the Fourth Industrial Revolution becomes a reality, the working environment undergoes a radical transformation and human work has undergone transformation around new concepts (Braña, 2019; Rosak-Szyrocka, 2025b; Rosak-Szyrocka & Wolniak, 2025a). The thorough integration of cyber-physical systems, IoT systems, AI, real-time data analytics, and ML has positioned workplace safety production ecosystems (Mourtzis et al., 2023; Nikolakis et al., 2019; Yang et al., 2024).

The traditional view based on control, hierarchy, and reactivity is no longer adequate in this context. Modern defence involves the application of systemic methodology in which technology, humanism, and social features are interwoven.

Human-centric safety can be seen as a paradigm shift between Industry 4.0 and Industry 5.0. Whereas in the case of Industry 4.0, automation, efficiency, and data integration were on top of the agenda; for Industry 5.0, increasing humanistic values, sustainable development, and employee well-being should take centre stage (Gamberini & Pluchino, 2024; Pasman & Behie, 2024; Rosak-Szyrocka & Knop, 2024). From this standpoint, safety is no longer a pure technical issue of control, but rather it is the result of a value co-creation process in which humans, organizations, and technologies interact ethically, transparently, and sustainably (Ávila-Gutiérrez et al., 2022).

In the human-centred model, the human is an active agent and adaptable towards safety. Rather than sidelining human expertise, modern systems capitalize on humans' capacity to learn; their ability to respond with robots is better in cases of uncertainty and in putting data into context. In this context, human-in-the-loop (HITL) (Iyenghar, 2025; Rosak-Szyrocka & Wolniak, 2025a; Rosak-Szyrocka & Wolniak, 2025b) and human-on-the-loop (HOTL) (Okola et al., 2025) dimensions have already been identified as key topics that refer to different levels of human intervention in decision-making tasks in AI-driven systems.

In the HITL scenario, a defined operator is actively making decisions – assisted by the AI system's mechanism of data analysis and help – yet it is still ultimately their decision. This model is used, for example, in an automatic production system where human acceptance is required before AI can execute its action. By contrast, in the HOTL paradigm, the human is supervising (overseeing) autonomous activities, serving as a fail-safe to prevent errors or ethical transgressions (Jiang et al., 2025; Okola et al., 2025). HOTL result is a change from the role of monitor operator towards that of co-decision maker and ethical stewardship in the control system, which goes to the heart of Industry 5.0 (Abouhawwash et al., 2024; Gamberini & Pluchino, 2024; Rame et al., 2024). Table 1.3. presents human–technology interaction models with respect to human-centred safety.

The emergence of Industry 4.0 and the subsequent development of the paradigm of Industry 5.0 have revolutionized the human role within industrial systems from a mere executor of simple physical tasks to a co-creator – together with digital entities – of complex, automated, and ethically aligned safety-related systems. The perception of the operator's role in the stages of industrial evolution, their areas of responsibility, and relationships has changed significantly. At first, man was considered as an arm of the machine, a servant of its law. In Industry 5.0, however, the human becomes a technological and ethical partner in a co-responsible manner for safety, organizational values, and sustainable development (Islam et al., 2025; Trstenjak et al., 2025; Wang et al., 2023).

Current concepts of occupational safety have shifted from a reactive, control-oriented style towards systemic, predictive, and cooperative approaches. Here, operator tasks are not limited to the execution of commands or response to technical issues but also include direct involvement in the steps of decisions supported by AI, real-time data analysis, and ML processes. The operator's role is central to the development of this digital safety culture – one that moves from human–technology interaction based on trust, transparency, and shared responsibility (Jan Emory & Kippenbrock, 2025; Kiangala & Wang, 2025). Table 1.4 presents the transformation of the human function (operator) and the evolvement of the operator-centred concept from Operator 1.0 to Operator 5.0.

TABLE 1.3
Human–Technology Interaction Models with Respect to Human-Centred Safety

Model	Level of Human Involvement	Description	Examples	Human-Centred Dimension
HITL	High – the human is involved in all decision-making.	The human processes the AI system's output and makes the ultimate decision; the system offers cognitive support.	Predictive safety systems in factories (the person validates that AI conclusions to stop a production line are reasonable); self-driving industrial vehicles with hand-validated decisions.	Allows for the application of human context, intuition, and experience in technical decisions; supports individual accountability.
HOTL	Medium – human in the loop for everything, but not making every decision.	It is an autonomous system, and a human oversees its actions, with intervention in emergencies.	Human-supervised safety systems (e.g., considering the monitoring of collaborative robots and autonomous logistics systems).	Promotes ethical responsibility, encourages a well-defined trade-off between AI automation and human control, and increases the trust in human factors.

Source: Own study.

TABLE 1.4
Transformation of the Human Function (Operator) and the Evolvement of the Operator-Centred Concept from Operator 1.0 to Operator 5.0

Operator	Description	Examples
Operator 1.0	The operator repeats the manual motions. Labour is highly physical and relatively low in cognition; however, workers were heavily dependent on the pace of the machine.	The relationship between humans and technology is strictly auxiliary – the person is regarded as part of the system, subordinate to its functioning. Safety is conceived in a restricted physical sense, namely to some extent protection against mechanical injuries and accidents, which does not take account of organizational or psychological aspects.
Operator 2.0	The operator oversees the standardized production process and keeps electrically driven machinery in smooth running order. The emphasis is on productivity and equipment management.	This phase is characterized by the development of awareness of formal safety regulations and legal norms in occupational protection at work. Work safety and ergonomics first become relevant. Nevertheless, safety is reactive and compliance-oriented, designed to prevent errors rather than taking the human as part of the system.

(Continued)

TABLE 1.4 (CONTINUED)

Operator	Description	Examples
Operator 3.0	The operator possesses the skills of cognition and analysis and supervises automatic and computer-controlled equipment, realizing systems while it analyses the operational data.	With the emergence of integrated Safety Management Systems (SMS) and safety culture, the human factor plays an increasingly central role in system reliability. The operator is seen as part of the socio-technical system, capable of interpreting situations and learning from organizations. Safety moves from a compensatory to a systemic, learning focus.
Operator 4.0	The digital operator leverages real-time data, IoT, and AI capabilities to observe and analyse production systems. Decision-making is based on interpreting data and cooperating with autonomous entities.	The HITL model develops, in which humans actively engage in AI-supported decision-making. The operator collaborates with technology in the interactive digital environments, thereby setting a framework that is ethical and contextualized. The operator acts as an organizational guardian. A new digital safety culture emerges in which data-informed decisions are countered with human judgement, trust, and accountability.
Operator 5.0	The operator is a co-creator and the decision partner in the technological, ethical, and social dimensions of company performance. Their duty is not merely to supervise but also to take joint responsibility for values, security, and sustainability.	Operating within the HOTL framework, the human oversees autonomous decision systems and adds moral reasoning, empathy, and situational awareness. Safety becomes sustainable, technological, ethical, and ecological together. The human takes centre stage of a collaborative, value-based, and sustainable safety ecosystem.

Source: Own study based on Islam et al. (2025), Kiangala and Wang (2025), and Trstenjak et al. (2025).

1.3 MODERN SAFETY CULTURE IN AN ERA OF DATA AND AUTOMATION

Safety culture refers to a subset of the general organizational culture that comprises a set of values, norms, beliefs, and actions, which define how employees view hazards and act in relation to risks (Alabdaly et al., 2024; Kiptulon et al., 2024). In contemporary interpretation drawn from quality and management sciences, safety culture is a systemic property of high reliability organizations (HROs) that incorporate technical, procedural, social, and human components to form an integrated strategy for managing uncertainty (Jabłoński & Jabłoński, 2021; Murray et al., 2024).

The current culture of safety is being fundamentally changed by digitalization, automation, and the use of AI (Rosak-Szyrocka, 2024; Tripathi & Rosak-Szyrocka, 2024). In the era of Industry 4.0 and now 5.0, OSH is no longer a set of reactive rule-based compliance but rather an ongoing cognitive process, whereby humans, technology, and organizations combine to form an interconnected, adaptive system that is continuously learning (Digmayer & Jakobs, 2019; Javaid et al., 2022). In this way, safety is no longer just the outcome of adhering to standards, but an organization's ability to anticipate and adapt and make informed decisions in non-linear operational settings (Jabłoński & Jabłoński, 2021; Pandey et al., 2025; Shaheen et al., 2022).

The contemporary safety culture represents a major step in the progression of OHS management: from reactive, compliance-oriented models to predictive systems that are based on data, organizational learning, and shared responsibilities. As identified in recent literature, safety culture is an essential part of organizational culture and it's one of the key elements for systemic resilience where organizations are able to respond to complex, dynamic, and technologically mature design operational environments within Industry 4.0 and Industry 5.0 paradigms (Afonso-Fernandes et al., 2026; Haleem et al., 2025; Hollnagel et al., 2017).

Safety is no longer the compliance to a set of prescriptive norms, rules, and procedures. In contemporary organizations, it is defined as a cognitive activity in which humans, technology, and organizational structures interact to create information flows, interpret data, and learn. In this view, safety is not solely a factor that indicates compliance with rules; it also embodies an expression of maturity derived from reflexivity, adaptive potential, and collective learning (Doyle et al., 2025). There are four pillars of modern safety culture (Figure 1.1):

1. **Transparency of data and decisions** – modern safety systems are premised on openness and verifiability as the basis for making decisions. In automation and AI-driven environments, where decision-making is being done by the algorithms, it becomes hugely important that the algorithmic process should always be explainable and traceable to human judgement. Explainable artificial intelligence (XAI)[1] principles and auditable safety

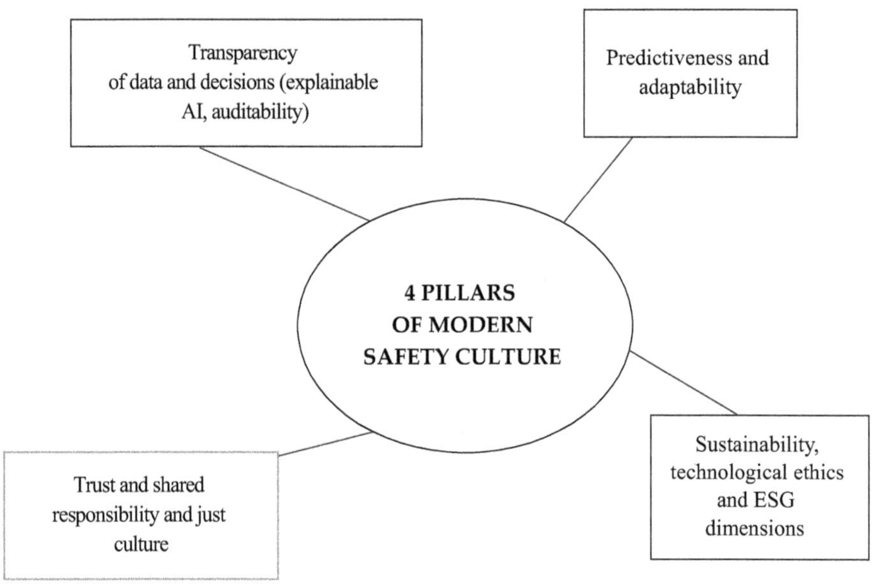

FIGURE 1.1 Four pillars of modern safety culture.

(Own study.)

architectures can be used in applying such approaches to have transparency in the automated decisions along with their social, ethical, legal, and operational consideration (Chen & Li, 2025; Kiruba & Sofia, 2025; Slil et al., 2025). This type of transparency builds trust between the parties and is key to evidence-based and responsible safety management (Wiratsin & Ragkhitwetsagul, 2025).

2. **Predictiveness and adaptability** – contemporary organizations are transitioning from reactive incident response towards predictive safety systems that leverage real-time information to detect and prevent probable dangers (Kabashkin et al., 2025). ML models process operator performance, sensor data, and cognitive load patterns to enable early warning systems and dynamic risk management regimes. IoT, digital twin, and AI emerge in the resilient organization that can continuously adapt to a dynamic operational environment (Pandey et al., 2025).
3. **Trust and shared responsibility** – modern safety culture is based on trust, transparency, and the philosophy of just culture[2] (Burns et al., 2006; Dekker, 2016; Walsh, 2019). It encourages free expression of errors, deviations from standard practice or 'near misses'[3] and uses them as sources for learning rather than reasons for punishment (Bachar et al., 2025; Muley et al., 2025; Wu et al., 2025). This will enhance a culture of psychological safety, and the reporting of risks creates an environment that is conducive to learning as an organization (Imran et al., 2025; Madsgaard & Svellingen, 2025; Montgomery et al., 2025). Shared vigilance means that safety is not the sole property of health and safety departments, but an organizational value shared by all involved parties in a system (Abouhawwash et al., 2024; Friganovic et al., 2025).
4. **Sustainability, Technological Ethics, and Environmental, Social, and Governance (ESG) Dimensions** – in the Industry 5.0 framework, safety extends beyond the safeguarding of human life and health to include responsibility for the environment, society, and ethics. Modern methodologies are considering the systems of safety management as being included in the wider framework related to Environmental, Social, and Governance (ESG), which is one of the most significant parts of a sustainable strategy for an organization (Xue et al., 2025; Yoon et al., 2025). The Environmental (E) dimension is about reducing the risk of environmental harm, using resources efficiently, and avoiding technological failures with ecological implications. The Social (S) subcomponent is about employee well-being, health, and safety, while also ensuring a fair culture with respect to the discussion of errors and findings in general. The governance (G) dimension encompasses transparency, ethical leadership, and adherence to the explainable AI (XAI), which aims to ensure that decisions in automated systems related to safety are monitorable and accountable (Chen & Li, 2025; Kiruba & Sofia, 2025; Wiratsin & Ragkhitwetsagul, 2025). Combining ESG with the principle of sustainable safety develops a sustainability concept for safety, with an approach that regards safety as a long-term moral and organizational value device, and which joins the technological innovation with ethical and social responsibility building (Abouhawwash et al., 2024; Yoon et al., 2025).

In order to show how these concepts are being applied on the ground, concrete examples of organizations that have moved to predictive and data-driven safety management after a history of reactive safety management are presented in the following case studies.

CASE STUDY 1 PROTEX AI – PROGRESSING A DATA-DRIVEN SAFETY CULTURE IN EUROPEAN PORT OPERATIONS

An example of how a data-driven and transformational safety culture can be implemented in practice is the use case Protex is working on with one of Europe's biggest logistics and shipping ports. The case study illustrates how AI and computer vision can reinvent occupational safety management in rapidly changing industrial environments.

The effort was an attempt to increase awareness of risk and to minimize the number of incidents and close calls – a near miss is something that could have but did not result in injury and refers to 'systemic weakness'. Protex AI announced an AI-based video analytics solution controlled from the existing surveillance system of the port. The system monitored operational areas 24/7 in real time and issued alarms when unsafe practices were detected, such as people entering the crane operating zone, unauthorized pedestrian access at entry points, and encroachment by ground workers moving rubber-tyred gantry (RTGs). The results were significant:

- 73% decrease in reported safety events,
- 415% increase in near-miss visibility,
- 24/7 risk condition monitoring and alerting.

The implementation transformed the organization's safety model from reactive into proactive and adaptive. Early warning based on real-time data was possible in order to intervene specifically before events themselves occurred. The system enabled tailored safety campaigns to address high-risk practices via automated alerts and video feedback. Data from the system was reviewed weekly during a Project Safety Meeting to identify trends and implement corrective actions.

In addition to technological development, the implementation success depended largely on organizational and cultural change. To build trust in the new AI safety system, open communication and workforce engagement were key. Training sessions and clear messages have assuaged concerns about surveillance and job security and that the technology was there to support, rather than undermine human judgement.

This engaged process embodied the tenets of a just culture, driven by learning and collective accountability rather than retribution. Staff were motivated to report unsafe acts and near-miss incidents that had become important learning events

(Protex AI, 2024)

CASE STUDY 2 AI-BASED SAFETY TRANSFORMATION AT LOGISTICS CENTRES – THE SURVEILY WAY

AI and computer vision are becoming more and more integrated within industry, leading to a reshaping of the way modern organizations approach the management of OSH. One example of this conversion is the deployment of the Surveily AI-powered risk management system within a regional network of large European logistics and distribution centres.

This initiative was motivated by the need to improve workplace conditions in a dynamic high-intensity environment with frequent human–machine interactions and a significant potential for collisions in low-visibility areas. Before the implementation, the company's safety management system was reactive – there was a larger focus on analysing incidents after they had already occurred, rather than preventing them before they happened.

To meet this requirement, the company implemented an AI-based safety monitoring and analytics platform that combines information from surveillance cameras, IoT devices, and traffic management systems. The system's computer vision algorithms were constantly analysing working zones to automatically identify potentially dangerous activities, such as:

- distance between workers and active forklifts and transport vehicles,
- personal protective equipment absent or not used correctly,
- unauthorized access in the secured areas, and
- unsafe pedestrian movements in the load and unload zones.

When events were detected, they were reported to safety teams so that intervention could be taken in real time and risks proactively addressed. Data collected was also used as input for predictive trend analysis and weekly safety meetings (toolbox talks), where teams reviewed incident trends and discussed corrective actions. The feedback revealed tangible results from AI for safety transformation:

- 62% reduction in safety incidents,
- 1498% increase in near-miss visibility,
- violation of their control area was reduced by 85% (Surveily, 2025).

A crucial part of that was clear communication with the staff. The company stressed that AI was not a surveillance tool but a safety support, something to aid but not replace human judgement. This approach builds trust in the staff and reinforces the elements of just culture.

NOTES

1 Explainable artificial intelligence (XAI) is a principle of designing artificial intelligence algorithms in a manner that the decisions made by these algorithms are explainable and interpretable to human beings. The use of XAI for safety management systems promotes the traceability, audit ability, and ethical accountability of automated decision-making processes and enhances confidence between humans and AI-driven technologies.

2 Just Culture is the theory that human errors are a normal part of complex sociotechnical systems and not reasons for punishment. In today's organization, a just culture among all staff encourages trust, transparency, and open communication which are designed to promote the reporting of unsafe acts and near misses without fear of recrimination. Studies have demonstrated that the application of just culture principles greatly enhances employee engagement and hazard reporting and has directly led to a measurable reduction in workplace incidents.

3 Near misses are a basic predictive and learning tool in contemporary safety-focused organizations. Refer to events which could have led to an accident but were made safe by human action or system defences. Within contemporary safety management, they represent an essential source of information on underlying system issues and can be seen as the foundation for organizational learning and predictive safety functions.

REFERENCES

Abouhawwash, M., Rosak-Szyrocka, J., & Gupta, S. K. (2024). *Aspects of Quality Management in Value Creating in the Industry 5.0 Way*. CRC Press.

Afonso-Fernandes, J., Barbosa, J., Arezes, P., Pardo-Ferreira, C., Rubio-Romero, J. C., & Rodrigues, M. A. (2026). Assessing resilience potentials in management of occupational safety and health in hospitals: Development and validation of a tool. *Journal of Safety Science and Resilience*, *7*(1), 100247.

Airports Council International (ACI). (2016). *Safety Management Systems Handbook* (1st ed.). ACI World.

Alabdaly, A., Hinchcliff, R., Debono, D., & Hor, S.-Y. (2024). Relationship between patient safety culture and patient experience in hospital settings: A scoping review. *BMC Health Services Research*, *24*(1), 906. https://doi.org/10.1186/s12913-024-11329-w

Aven, T. (2022). A risk science perspective on the discussion concerning safety I, safety II and safety III. *Reliability Engineering & System Safety*, *217*, 108077.

Ávila-Gutiérrez, M. J., Suarez-Fernandez de Miranda, S., & Aguayo-González, F. (2022). Occupational safety and health 5.0—A model for multilevel strategic deployment aligned with the sustainable development goals of agenda 2030. *Sustainability*, *14*(11), 6741.

Bachar, R., Urlainis, A., Wang, K.-C., & Shohet, I. M. (2025). Optimal allocation of safety resources in small and medium construction enterprises. *Safety Science*, *181*, 106680.

Bartulović, D. (2021). Predictive safety management system development. *Transactions on Maritime Science*, *10*(1), 135–146. https://hrcak.srce.hr/clanak/375168

Bennouna, A., Boughaba, A., Djabou, S., & Mouda, M. (2025). Enhancing workplace well-being: Unveiling the dynamics of leader–member exchange and worker safety behavior through psychological safety and job satisfaction. *Safety and Health at Work*, *16*(1), 36–45.

Berx, N., Decré, W., Schutter, J. de, & Pintelon, L. (2025). A harmonious synergy between robotic performance and well-being in human-robot collaboration: A vision and key recommendations. *Annual Reviews in Control*, *59*, 100984.

Boto-Álvarez, A., & García-Fernández, R. (2020). Implementation of the 2030 agenda sustainable development goals in Spain. *Sustainability*, *12*(6), 2546. https://www.mdpi.com/2071-1050/12/6/2546

Braña, F.-J. (2019). A fourth industrial revolution? Digital transformation, labor and work organization: A view from Spain. *Journal of Industrial and Business Economics*, *46*(3), 415–430. https://doi.org/10.1007/s40812-019-00122-0

Burns, C., Mearns, K., & McGeorge, P. (2006). Explicit and implicit trust within safety culture. *Risk Analysis: An Official Publication of the Society for Risk Analysis*, *26*(5), 1139–1150.

Chen, Q., & Li, B. (2025). Explainable artificial intelligence (XAI)-driven probabilistic image-based structural health monitoring of reinforced concrete beams with shear reinforcements. *Automation in Construction*, *180*, 106549.

Chia, G., Lim, S. M., Sng, G. K. J., Hwang, Y.-F. J., & Chia, K. S. (2019). Need for a new workplace safety and health (WSH) strategy for the fourth industrial revolution. *American Journal of Industrial Medicine*, *62*(4), 275–281.

Dekker, S. (2016). *Just Culture*. CRC Press.

Digmayer, C., & Jakobs, E.-M. (2019). Developing safety cultures for Industry 4.0. New challenges for professional communication. In *2019 IEEE International Professional Communication Conference (ProComm)* (pp. 218–225). IEEE.

Doyle, K., Moore, K., & Foster, P. (2025). Criteria-driven socio-environmental maturity modelling for mining: Driving positive sustainability attitudes and perceptions at diverse operational scales. *Resources Policy*, *105*, 105606.

Eurostat. (2024, November 8). *Accidents At Work – Number of Fatal Accidents by Economic Activity (HSW_N2_02)* [Data set]. Statistical Office of the European Union. https://ec.europa.eu/eurostat/databrowser/view/hsw_n2_02/default/table?lang=en

Fink, D. (2010). Road Safety 2.0: Insights and implications for government. In M. Drobnič, G. Jereb, T. Rajkovič, & V. Bobek (Eds.), *Proceedings of the 23rd Bled eConference: eTrust: Implications for the Individual, Enterprises and Society* (pp. 274–285). Bled, Slovenia, June 20–23, 2010. Association for Information Systems. https://aisel.aisnet.org/bled2010/8

Friganovic, A., Boskovic, S., Krupa Nurcek, S., Kovacevic, I., Kosydar-Bochenek, J., & Filipovic, B. (2025). Editorial: Patient and medical staff safety and healthy work environment in the 21st century. *Frontiers in Public Health*, *13*, 1677117.

Gamberini, L., & Pluchino, P. (2024). Industry 5.0: A comprehensive insight into the future of work, social sustainability, sustainable development, and career. *Australian Journal of Career Development*, *33*(1), 5–14.

Haleem, A., Javaid, M., & Singh, R. P. (2025). Encouraging Safety 4.0 to enhance industrial culture: An extensive study of its technologies, roles, and challenges. *Green Technologies and Sustainability*, *3*(3), 100158.

Hollnagel, E. (2018). *Safety–I and Safety–II*. CRC Press.

Hollnagel, E., Woods, D. D., & Leveson, N. (2017). *Resilience Engineering: Concepts and Precepts*. CRC Press.

Imran, M., Li, J., Bano, S., & Rashid, W. (2025). Impact of democratic leadership on employee innovative behavior with mediating role of psychological safety and creative potential. *Sustainability*, *17*(5), 1879. https://www.mdpi.com/2071-1050/17/5/1879

International Civil Aviation Organization. (2013). *Safety Management Manual (SMM)* (3rd ed., Doc 9859). ICAO.

Islam, M. T., Sepanloo, K., Woo, S., Woo, S. H., & Son, Y.-J. (2025). A review of the industry 4.0 to 5.0 transition: Exploring the intersection, challenges, and opportunities of technology and human–machine collaboration. *Machines*, *13*(4), 267. https://www.mdpi.com/2075-1702/13/4/267

Iyenghar, P. (2025). Clever Hans in the loop? A critical examination of ChatGPT in a human-in-the-loop framework for machinery functional safety risk analysis. *Eng*, 6(2), 31. https://www.mdpi.com/2673-4117/6/2/31

Jabłoński, M., & Jabłoński, A. (2021). Shaping the safety culture of high reliability organizations through digital transformation. *Energies*, *14*(16), 4721. https://www.mdpi.com/1996-1073/14/16/4721

Jan Emory, D., & Kippenbrock, T. A. (2025). Patient safety ratings of hospitals by registered nurses, managers, and executives: A retrospective analysis of patient safety culture data. *Nursing Outlook*, *73*(6), 102587.

Javaid, M., Haleem, A., Singh, R. P., Rab, S., & Suman, R. (2022). Exploring impact and features of machine vision for progressive industry 4.0 culture. *Sensors International*, *3*, 100132.

Jiang, J., Murray, D., Stankovic, V., Stankovic, L., Hibert, C., Pytharouli, S., & Malet, J.-P. (2025). A human-on-the-loop approach for labelling seismic recordings from landslide site via a multi-class deep-learning based classification model. *Science of Remote Sensing*, *11*, 100189.

Kabashkin, I., Fedorov, R., & Perekrestov, V. (2025). Decision-making framework for aviation safety in predictive maintenance strategies. *Applied Sciences*, *15*(3), 1626. https://www.mdpi.com/2076-3417/15/3/1626

Kiangala, K. S., & Wang, Z. (2025). A generative pre-trained transformer industrial bot to improve operators' working experience in a small Industry 5.0 factory. *The International Journal of Advanced Manufacturing Technology*, *136*(7–8), 3525–3541. https://doi.org/10.1007/s00170-025-15033-9

Kiptulon, E. K., Elmadani, M., Limungi, G. M., Simon, K., Tóth, L., Horvath, E., et al. (2024). Transforming nursing work environments: The impact of organizational culture on work-related stress among nurses: A systematic review. *BMC Health Services Research*, *24*(1), 1526. https://doi.org/10.1186/s12913-024-12003-x

Kiruba, P., & Sofia, D. (2025). An explainable artificial intelligence (XAI) based driver fatigue detection system for safe and care human life (SCHL). *International Journal of Environmental Sciences*, *11*(8s), 839–855. http://theaspd.com/index.php/ijes/article/view/895

Kumar, N., Pal, S. K., Agarwal, P., Rosak-Szyrocka, J., & Jain, V. (2024). *Human-Machine Collaboration and Emotional Intelligence in Industry 5.0*. IGI Global.

Lemos, J., Souza, V. B. de, Falcetta, F. S., Almeida, F. K. de, Lima, T. M., & Gaspar, P. D. (2024). Enhancing workplace safety through personalized environmental risk assessment: An AI-driven approach in Industry 5.0. *Computers*, *13*(5), 120. https://www.mdpi.com/2073-431x/13/5/120

Leveson, N. G. (2016). *Engineering a Safer World*. The MIT Press.

Liu, K., Kou, Y., Liu, Y., & Yang, X. (2024). Relationship network of safety management elements in the construction industry under the perspective of resilience. *Engineering, Construction and Architectural Management*.

Liu, Z., Xie, K., Li, L., & Chen, Y. (2020). A paradigm of safety management in industry 4.0. *Systems Research and Behavioral Science*, *37*(4), 632–645.

Madsgaard, A., & Svellingen, A. (2025). The benefits and boundaries of psychological safety in simulation-based education: An integrative review. *BMC Nursing*, *24*(1), 922. https://doi.org/10.1186/s12912-025-03575-y

Maia, E., Wannous, S., Dias, T., Praça, I., & Faria, A. (2022). Holistic security and safety for factories of the future. *Sensors*, *22*(24), 9915. https://www.mdpi.com/1424-8220/22/24/9915

McCaffrey, L., deBoer, J., Vogelsang, M., Giné-Garriga, M., McCann, B., & Dall, P. (2025). A comparison of office workers' co-creation experiences: A health CASCADE multiple case study. *Public Health*, *241*, 75–82.

Min, J., Kim, Y., Lee, S., Jang, T.-W., Kim, I., & Song, J. (2019). The fourth industrial revolution and its impact on occupational health and safety, worker's compensation and labor conditions. *Safety and Health at Work*, *10*(4), 400–408.

Mohy, A. A., Bassioni, H. A., Elgendi, E. O., & Hassan, T. M. (2024). Innovations in safety management for construction sites: The role of deep learning and computer vision techniques. *Construction Innovation*.

Montgomery, A., Chalili, V., Lainidi, O., Mouratidis, C., Maliousis, I., Paitaridou, K., & Leary, A. (2025). Psychological safety and patient safety: A systematic and narrative review. *PLoS One*, *20*(4), e0322215. https://doi.org/10.1371/journal.pone.0322215

Mourtzis, D., Angelopoulos, J., & Panopoulos, N. (2023). The future of the human–machine Interface (HMI) in society 5.0. *Future Internet*, *15*(5), 162. https://www.mdpi.com/1999-5903/15/5/162

Muley, S., Wang, C., Aghazadeh, F., & Bangaru, S. S. (2025). Assessing construction near-miss detection proficiency for workers under stressor conditions using psychophysiological measures: An eye-tracking investigation. *Applied Sciences*, *15*(3), 1558. https://www.mdpi.com/2076-3417/15/3/1558

Murray, J. S., Baghdadi, A., Dannenberg, W., Crews, P., & Walsh, N. D. (2024). The role of high reliability organization foundational practices in building a culture of safety. *Federal Practitioner: For the Health Care Professionals of the VA, DoD, and PHS*, *41*(7), 214–221. https://pmc.ncbi.nlm.nih.gov/articles/pmc11473027/

Musarat, M. A., Alaloul, W. S., Irfan, M., Sreenivasan, P., & Rabbani, M. B. A. (2023). Health and safety improvement through industrial revolution 4.0: Malaysian construction industry case. *Sustainability*, *15*(1), 201. https://www.mdpi.com/2071-1050/15/1/201

Nikolakis, N., Maratos, V., & Makris, S. (2019). A cyber physical system (CPS) approach for safe human-robot collaboration in a shared workplace. *Robotics and Computer-Integrated Manufacturing*, *56*, 233–243.

Okola, I., Ochieng, D. O., Miriti, E. K., & Ong'isa, G. O. (2025). The human-on-the-loop evaluation of multiobjective optimization algorithms for solving a real-world problem that integrates the food-energy-water nexus security and climate change vulnerability. *Results in Control and Optimization*, *20*, 100606.

Pandey, S., Singh, A. K., & Parhi, S. (2025). Transforming workplace safety through leveraging predictive analytics and explainable AI in steel industries. *Proceedings of the Institution of Mechanical Engineers, Part O: Journal of Risk and Reliability*, *239*(6), 1332–1345.

Pasman, H. J., & Behie, S. W. (2024). The evolution to industry 5.0/Safety 5.0, the developments in society, and implications for industry management. *Journal of Safety and Sustainability*, *1*(4), 202–211.

Protex AI. (2024, May 23). *415% Increase in near miss visibility: AI-driven port safety* [Case study]. https://www.protex.ai/case-studies/73-decrease-in-incidents-ai-driven-port-safety-by-protex.ai

Rame, R., Purwanto, P., & Sudarno, S. (2024). Industry 5.0 and sustainability: An overview of emerging trends and challenges for a green future. *Innovation and Green Development*, *3*(4), 100173.

Ramírez-Peña, M., Cerezo-Narváez, A., Otero-Mateo, M., Pastor-Fernández, A., & Batista, M. (2025). AI-enhanced safety in project-based production: A new era in workplace risk. *Artificial Intelligence and Digital Transformation*, *78*, 43–57. https://doi.org/10.1007/978-3-032-00118-4_4

Rodriguez, A., Douphrate, D. I., Pineda Reyes, A. L., Zavala, S., Cabrera Cruz, A. V., Alegria, M. D., et al. (2025). Promoting workplace health, safety, and well-being among essential agricultural workers through vaccine-preventable infectious diseases training in the Rio Grande Valley. *Journal of Agromedicine*, *30*(1), 57–67.

Rosak-Szyrocka, J. (2024). The era of digitalization in education where do universities 4.0 go? *Management Systems in Production Engineering*, *32*(1), 54–66.

Rosak-Szyrocka, J. (2025a). Engineering the future of higher education: A VOSviewer analysis of smart university trends in the digitalization and industry 5.0 era. *Management Systems in Production Engineering*, *33*(1), 8–23.

Rosak-Szyrocka, J. (2025b). Reskilling and upskilling in the age of AI: A practical guide to workforce transformation. In *Artificial Intelligence, Machine Learning, Data Analytics and Automation for Business Management Series* (1st ed.). CRC Press LLC.

Rosak-Szyrocka, J., Gupta, S. K., Shahbaz, M., & Launer, M. (2025). *Enhancing Smart Universities with Emotional Intelligence: Competences for Digital Age Learning*. Taylor & Francis.

Rosak-Szyrocka, J., & Knop, K. (2024). Sustainable development universities-opportunities & challenges on the road to society 5.0 from the key stakeholder perspective. *PLoS One*, *19*(8), e0308929. https://doi.org/10.1371/journal.pone.0308929

Rosak-Szyrocka, J., & Tiwari, S. (2023). Structural equation Modeling (SEM) to test sustainable development in university 4.0 in the ultra-smart society era. *Sustainability*, *15*(23), 16167. https://www.mdpi.com/2071-1050/15/23/16167

Rosak-Szyrocka, J., & Wolniak, R. (2025a). *AI-Driven Sustainability: The Future of Human Resources Management/Joanna Rosak-Szyrocka, Radoslaw Wolniak*. CRC Press.

Rosak-Szyrocka, J., & Wolniak, R. (2025b). *Smart Universities in Smart Cities: Shaping the Future of Education and Urban Innovation*. Taylor & Francis.

Rosak-Szyrocka, J., Zywiolek, J., Kulinska, E., & Matulewski, M. (2021). Analysis of enterprises' readiness in for industry 4.0 implementation: The case of Poland. *European Research Studies Journal*, *XXIV*(3), 615–628. https://www.um.edu.mt/library/oar/handle/123456789/83208

Rosak-Szyrocka, J., Zywiołek, J., Nayyar, A., & Naved, M. (2023a). *The Role of Sustainability and Artificial Intelligence in Education Improvement*. CRC Press.

Rosak-Szyrocka, J., Zywiołek, J., & Shahbaz, M. (2023b). *Quality Management, Value Creation, and the Digital Economy*. Routledge.

Samost-Williams, A., & Brook, K. (2025). Using safety-I, II, and III to level up patient safety. *Current Opinion in Anaesthesiology*, *38*(6), 741–747.

Sarvari, H., Edwards, D. J., Rillie, I., & Posillico, J. J. (2024). Building a safer future: Analysis of studies on safety I and safety II in the construction industry. *Safety Science*, *178*, 106621.

Scandelai, N. T. (2025). Technological advancements in occupational health: Enhancing workplace safety and well-being. *Brazilian Journal of Development*, *11*(3), e78096. https://ojs.brazilianjournals.com.br/ojs/index.php/brjd/article/view/78096

Shaheen, F., Lodhi, M. S., Rosak-Szyrocka, J., Zaman, K., Awan, U., Asif, M., et al. (2022). Cleaner technology and natural resource management: An environmental sustainability perspective from China. *Clean Technologies*, *4*(3), 584–606. https://www.mdpi.com/2571-8797/4/3/36

Slil, E., Iyiola, K., Alzubi, A., & Aljuhmani, H. Y. (2025). Impact of safety leadership and employee morale on safety performance: The moderating role of harmonious safety passion. *Buildings*, *15*(2), 186. https://www.mdpi.com/2075-5309/15/2/186

Surveily. (2025). *Enhancing Safety and Visibility in Logistics: Case Study*. https://surveily.com/case-studies/logistics

Szopik-Depczyńska, K., Kędzierska-Szczepaniak, A., Szczepaniak, K., Cheba, K., Gajda, W., & Ioppolo, G. (2018). Innovation in sustainable development: An investigation of the EU context using 2030 agenda indicators. *Land Use Policy*, *79*, 251–262.

Tripathi, S., & Rosak-Szyrocka, J. (2024). *Impact of Artificial Intelligence on Society*. CRC Press.

Trstenjak, M., Benešova, A., Opetuk, T., & Cajner, H. (2025). Human factors and ergonomics in industry 5.0—A systematic literature review. *Applied Sciences*, *15*(4), 2123. https://www.mdpi.com/2076-3417/15/4/2123

Walsh, P. (2019). What is a 'just culture'? *Journal of Patient Safety and Risk Management*, *24*(1), 5–6.

Wang, H., Lv, L., Li, X., Li, H., Leng, J., Zhang, Y., et al. (2023). A safety management approach for industry 5.0's human-centered manufacturing based on digital twin. *Journal of Manufacturing Systems*, *66*, 1–12.

WHO & ILO. (2021, September 17). *WHO/ILO: Almost 2 Million People Die from Work-Related Causes Each Year*. World Health Organization. https://www.who.int/news/item/17-09-2021-who-ilo-almost-2-million-people-die-from-work-related-causes-each-year

Wiratsin, I.-O., & Ragkhitwetsagul, C. (2025). Effectiveness of explainable artificial intelligence (XAI) techniques for improving human Trust in Machine Learning Models: A systematic literature review. *IEEE Access*, *13*, 121326–121350.

Wu, Z., Wang, S., Li, L., Wang, Y., & Wang, X. (2025). Ship near-miss risk identification method based on multi-modal deep fusion model under complex weather and sea conditions. *Ocean Engineering*, *329*, 121119.

Xue, Y., Zhang, C., Mentel, G., Mohammed, K. S., Rosak-Szyrocka, J., & Zhao, X. (2025). Impact of ESG rating divergence on corporate resilience: Insights from China's A-share market. *Environment, Development and Sustainability*, 1–26. https://doi.org/10.1007/s10668-025-06799-9

Yang, X., Xing, Q., Yang, J., Yao, Y., Tian, K., & Cai, H. (2024). Research on influencing factors and mechanisms of human-machine safety collaboration behavior in coal mines based on DEMATEL-ISM. *SAGE Open*, *14*(1). https://doi.org/10.1177/21582440241239126

Yitmen, I., Almusaed, A., & Alizadehsalehi, S. (2024). Facilitating construction 5.0 for smart, sustainable and resilient buildings: Opportunities and challenges for implementation. *Smart and Sustainable Built Environment*, *15*(2), 729–764. https://doi.org/10.1108/SASBE-04-2024-0127

Yoon, C., Kwak, H., & Jung, S. (2025). A study on ESG evaluation indicators through chemical accident data analysis and double materiality assessment. *PRO*, *13*(2), 323. https://www.mdpi.com/2227-9717/13/2/323

2 Key AI Technologies for Safety System

2.1 ML AND AD ALGORITHMS

Nowadays, an emerging trend in the design of advanced safety systems is machine learning (ML) methods that can be used to analyse process data, uncovering hidden patterns and enabling early risk detection. ML is the computer-based approach that allows the raw industrial data – from sensors, loggers, SCADA systems, or digital platforms – to be transformed into operational knowledge that makes it possible to predict failures, detect deviations early on, and make decisions based on information. Security is an area that can benefit quite a lot from ML models, since we can move from reactive responses to preventive and proactive actions (Vivian et al., 2025; Wei et al., 2023).

The four most common classes of ML approaches can be found in the literature related to safety (Figure 2.1): supervised learning, unsupervised learning, semi-supervised learning, and reinforcement learning (Dong et al., 2025; Guo et al., 2021; Razzaq & Shah, 2025). All of these paradigms serve to process different data and consequently address the specific needs of industry-scale risk monitoring and anomaly detection.

In supervised learning, data are labelled thanks to a ground truth (e.g., 'functional' vs 'faulty'), and the models are trained on these labelled datasets (Cherry & Qian, 2025; Da Silva et al., 2025; Guan et al., 2025). Especially in the presence of historical data that explains normal operations as well as categories of known faults, this inventory can be used to achieve high effectiveness. It has been extensively used in failure prediction, fault diagnosis, machine state classification, and process phase detection. Quatrini et al. (2020) presented a pragmatic approach to an anomaly detection system and process phase classifier based on supervised ML classifiers, which was proven beneficial for safety and maintenance decision support.

On the contrary, unsupervised learning is used when labelled failure data are extremely rare (a problem that is common on safety-critical systems, where incidents hardly ever happen). Such models analyse the data representation searching for clusters, hidden patterns, or deviations from the acquired profile of normal operation. Systematic reviews of anomaly detection techniques show that clustering, dimensionality reduction, and autoencoders are very popular in industry, as they are able to detect unseen anomalies (Grunova et al., 2024; Nassif et al., 2021).

The literature has more recently highlighted the need for semi-supervised learning, an intermediate paradigm that exploits a small set of labelled samples and a large amount of unlabelled data. In fact, this is very important for industrial safety because there are a few labelled failures but a huge amount of unlabelled no-failure data (Gan et al., 2018; Gan et al., 2020; Gan et al., 2023). Semi-supervised models leverage the

DOI: 10.1201/9781003754213-3

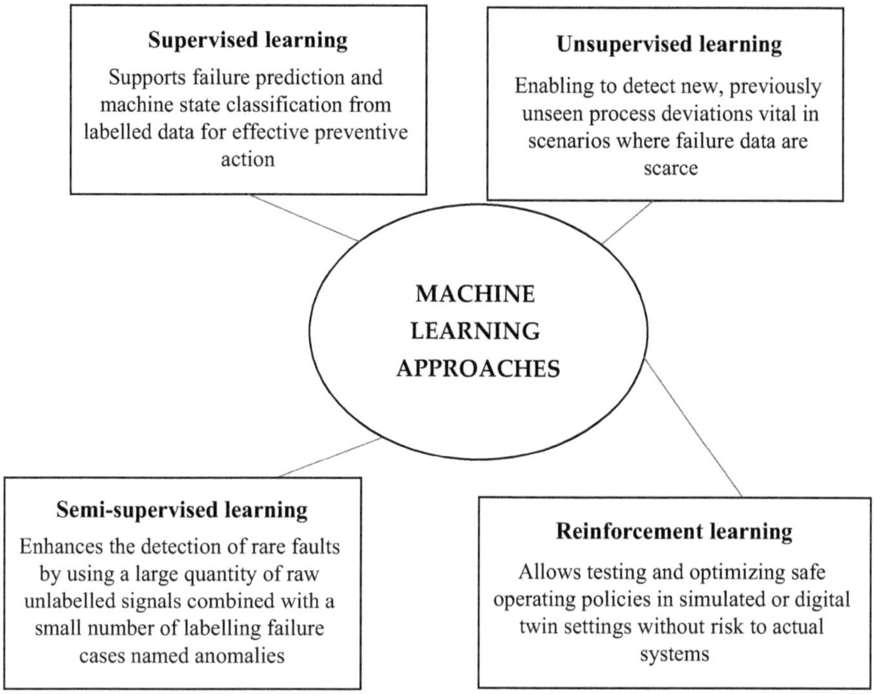

FIGURE 2.1 ML approaches in industry safety.

(Own study.)

structure of unlabelled data to enhance predictive performance, outperforming classical supervised approaches when there is only a small proportion of abnormal cases with labels.

Recent research has demonstrated that a semi-supervised learning paradigm is well suited to real-world anomaly detection problems in industry, which often possess limited labelled failure examples to work with, yet the models trained on predominantly normal operating patterns tend to detect deviations more effectively than fully supervised methods (Neupane et al., 2024; Serradilla et al., 2021; Zhang, 2025).

Reinforcement learning (RL) is a paradigm that enables a system to learn over time which decisions are optimal based on outcomes of its actions through trial and error (Su et al., 2025). At each iteration, the algorithm receives feedback from the environment and learns over time which strategies result in better operational outcomes (Reinforcement learning: An introduction, 1998).

In the literature on safety and reliability control, attention is increasingly focused on safe RLs, emphasizing the need to ensure that safety constraints and risks are considered in systems that learn autonomously (Inamdar et al., 2024; Wabersich & Zeilinger, 2021). Research has shown that RL is especially promising when applied in digital-twin-based environments, which offer a high-fidelity simulation of industrial processes. With such controlled environments, it is feasible to test operational conditions that might be costly, challenging, or dangerous to test in real systems

(Francis et al., 2024). Thus, RL can enhance operating procedures, response plans, and control laws related to safety performance, particularly for issues involving uncertainty, disturbances, or time variations (Khdoudi et al., 2024).

The function of RL is to support decision-making by enabling informed and safe operating procedures. Thus, within the safety-management domain, RL can be used to improve organizational resilience if implemented in environments that support the controlled simulation and evaluation of critical or dangerous situations (Khdoudi et al., 2024). Table 2.1 presents a comparison of ML approaches.

TABLE 2.1
Comparison of ML Approaches

ML Approach	Definition	Examples of Algorithms	Application in Safety
Supervised learning (SL)	Models are trained to map between input features and output labels. Effective when historical data banks are available that describe both normal and fault conditions. Extensively adopted in industrial condition monitoring and failure prediction.	SVM, Random Forest, Logistic Regression, MLP	Fault diagnosis, failure prediction, machine condition monitoring, process phase classification.
Unsupervised learning (UL)	Models process unsupervised data to find clusters, hidden structures, or anomalies out of the learnt image of normal behaviour. This is especially important in safety-critical domains where failures are infrequent.	K-means, DBSCAN, PCA, Autoencoders	Anomaly Detection for previously unknown anomalies, monitoring of process deviations, and cyber-anomalies in SCADA Networks.
Semi-supervised learning (SSL)	Integrates a tiny labelled dataset with extensive unlabelled datasets. SSL does enhance the prediction accuracy, in particular when labelled anomalies are scarce, by exploiting the structure of raw data.	Graph-based SSL, Autoencoders, Pseudo-labelling, HELM	Detection of small anomalous patterns, recognition of shallow degrading processes, and tracking industrial processes exhibiting big variability.
Reinforcement learning (RL)	An agent acts in an environment and learns to get more rewards.	Q-learning, SARSA, Deep Q-Network (DQN)	Testing dangerous scenarios, optimization of operational procedures, design of adaptive safety rules and policies, verification of emergency measures.

Source: Own study base on: Mitchell (2010), Nassif et al. (2021), Quatrini et al. (2020), Zhu et al. (2025).

Apart from classical ML approaches, anomaly detection (AD) methods are essential in modern safety systems (Yuan, 2025; Zhang et al., 2025b). They aim to identify anomalies in regular operation behaviour as well as early indicators of risk. In the industry, process operational threats appear mostly in subtle changes of monitored data that can be difficult to notice; thus, AD is the basis for an early warning system (Nassif et al., 2021).

The literature highlights that the success of proactive safety procedures is very much dependent on our capacity to identify long-term faults, so-called slow-developing failure states, i.e., before they are seriously incurred. The AD models, especially unsupervised learning-based models, can better capture these early indicators than the conventional threshold-based approach or visual assessment by a human operator (Grunova et al., 2024).

According to systematic reviews, industry applications mainly adopt a few dominant families of AD algorithms that are the main core of modern safety systems (Nassif et al., 2021; Wei et al., 2023). The literature characterizes AD as one of the most powerful mechanisms to identify new, unknown anomalies – this is especially important in safety-critical environments where incidents are extremely rare, and failure samples are scarce. These analyses also show that there are four main classes of AD algorithms used in industrial environments (Nassif et al., 2021; Wei et al., 2023):

- **Clustering-based approaches**: These are the methods that involve dividing data into groups where the entities within these groups share similar characteristics. Outlying observations from the centres of clusters are treated as outliers. Techniques like k-means or DBSCAN models are very popular for their strength to detect anomalous patterns, especially in non-labelled data (Li et al., 2021; Pu et al., 2021; Sánchez Vinces et al., 2025).
- **Dimensionality reduction techniques**: algorithms such as PCA or deep autoencoders learn succinct representations of a dataset. Any deviation from this expression is considered to be a risk. Autoencoders in particular have been increasingly used, as they are well suited to model complex industrial data distributions (Orlov et al., 2025; Serani & Diez, 2025; Wani, 2025).
- **Reconstruction-based approach**: These methods concentrate on the recovery of the input signal and track the reconstruction error (Peng et al., 2025; Qian et al., 2025). The increase in this error shows the abnormal operation state occurred. The review by Serradilla et al. (2021) demonstrates the efficacy of reconstruction-based techniques for machine diagnostics and predictive maintenance systems.
- **Rule-based and invariant-based approaches**: These techniques are commonly used in ICS/SCADA systems. They depend on rule-based rules of normal operational behaviour, and deviations from these rules are treated as potential threats. Such methods are especially useful when interpretability and auditability of decisions are needed (Ch et al., 2025; Zhu et al., 2025).

In the age of Industry 4.0, evidenced by complex, dynamic, and heterogeneous data streams, threshold-based rules are proved to be uncompetitive in outlier detection (Raptis et al., 2019; Sun et al., 2020). Several reviews point out that AD, which is based on learning the so-called normal behaviour profile of a system and detecting any deviation from it without any labelled failure data, plays a crucial role in emerging early warning systems and predictive safety analytics (Elía & Pagola, 2025; Kamat & Sugandhi, 2020; Kammerer et al., 2019).

AD algorithms are emerging in Industry 4.0 as enablers of modern safety supervision solutions due to the progressively more challenging character of industrial process data and the trend towards a proactive safety culture based on data analysis. AD methods are already explored in several strategic fields, such as predictive maintenance and advanced process monitoring (Kammerer et al., 2019), for the security of ICS/SCADA systems, allowing immediate recognition of anomalies and early warning signals forewarning hazardous events (Elía & Pagola, 2025; Kamat & Sugandhi, 2020).

(Elía & Pagola, 2025; Kamat & Sugandhi, 2020; Kammerer et al., 2019). Table 2.2 summarizes the main applications of AD algorithms in a modern industrial safety system.

TABLE 2.2
The Main Applications of AD Algorithms in a Modern Industrial Safety System

Application Area	Role of AD in Safety
Predictive Maintenance (PdM)	AD methods allow for early detection of very small evidence of wear and tear on machines by considering results downloaded from sensors monitoring, e.g., vibrations, temperature, acoustic patterns, or electrical parameters. These models can identify even very small variations from normal patterns of operation that would be hard or impossible for human operators to see in real time. AD also facilitates calculation of the Remaining Useful Life (RUL), allowing predictive monitoring rather than reactive and consequently reducing unplanned downtimes and maintenance costs.
Machine Condition Monitoring and Fault Diagnosis	In fault diagnosis, AD recognizes anomalous modes of operation, transients, and rare failure conditions that are unlabelled in training datasets. Through the analysis of time series of sequenced events and process signals, AD models are able to sample deviations from some modelled notions of normality. This leads to an early realization of potential slow-developing failures, which pose the greatest risk towards technical safety and operational continuity.
Process Safety	AD monitors critical process parameters, such as pressure, flow rates, temperatures, and chemical concentrations, for abnormal changes indicative of a process upset. The prompt detection of such abnormalities prevents these from developing into dangerous conditions, such as runaway reactions, excessive heating, leaks of hazardous substances, or debalancing processes. AD additionally enables the facilitated adherence to process safety guidelines through ongoing monitoring of process integrity.
ICS/OT/SCADA Cybersecurity	In industrial control systems, AD serves as a fundamental tool to identify cyber threats where traditional signature-based protection is not enough. AD can be used to monitor the exchange of network traffic, command sequences, and machine-to-machine communications to detect unusual/malicious activity. Such models can identify anomalies related to intrusions, control logic tampering, spoofing, privilege escalation, or integrity breaches. AD, therefore, furnishes a behaviour-based protective cushion that is indispensable in critical infrastructures.

(continued)

TABLE 2.2 (CONTINUED)

Application Area	Role of AD in Safety
Production Quality Control	AD algorithms process visual inspection data, metrology measurements, and sensor quality indicators to detect out-of-specification product or process patterns. This allows for identification of faults, such as microcracks, placement errors, surface defects, or dimensional non-uniformity. AD enables automated visual inspection, lowers levels of bad parts, and improves general production stability, particularly under high-variability conditions.
Environmental and Occupational Safety (HSE)	AD analyses IoT device, camera system, and localization sensor data to detect unsafe worker behaviours, unauthorized entry into hazardous zones, or dangerous human–machine interactions. These models also sense hazards in the environment, including air quality, temperature, or noise fluctuations. AD also provides early precautions to avoid accidents and enables real-time monitoring of HSE.
Energy and Utility Safety	AD recognizes alarming abnormal energy use, load profile, or stress signals in equipment and infrastructure, signalling overloads, fires, or network crashes that can be intercepted. Anomalies in the energy signature are identified as inefficient behaviour, a fault, or an undisclosed risk in the energy management system.
Digital-Twin-Based Safety Systems	In the context of digital twin environments, AD examines synthetic and simulated data that is derived from virtual models of industrial processes. This allows high-risk scenarios to be tested, AD models to be trained, and safety policies to be validated without real equipment being put at risk. With the integration of AD with digital twins, operational resiliency can be enhanced, and predictive safety planning would become more accurate.

Source: Own study base on: Chandola et al. (2009), Coelho et al. (2022), Raptis et al. (2019).

2.2 EDGE COMPUTING, INTERNET OF THINGS, AND REAL-TIME PROCESSING

Modern industrial companies face a context in which automation, variability, and the complexity of interactions among humans, machines, and digital elements continue to grow. These conditions fundamentally alter traditional methods of occupational safety because hazards in volatile environments can manifest within seconds and must be handled in near real time (Shah & Mishra, 2024; Tong et al., 2025). At the other end of the spectrum, it is also found in literature that traditional risk management models (from retrospective to predictive and up to post-event) do not work well under highly variable and short response times. In this respect, the use of artificial intelligence is a flagship to move from reactive through predictive to proactive safety, while ensuring policies of human-centred AI (Armenteros-Cosme et al., 2025; Fisher et al., 2023; Jetha et al., 2025).

One of the key premises of human-centred AI in safety is to design technology that helps people become more competent instead of simply replacing or reducing their role as decision-makers (Dutta & Mishra, 2024; Sánchez Vinces et al., 2025). Studies of human-AI collaboration show that technology acceptance, trust, and reduction in perceived surveillance or control are better supported when the system's design is user-centred, and employees are involved in designing and deploying solutions (Fiegler-Rudol et al., 2025; Kumar et al., 2024). In occupational safety, this results in systems with AI support for risk perception, advice, and organizational learning using data but leaving the ultimate protective decision to humans (El-Helaly, 2024; Montealegre-López, 2025).

A necessary condition for this safety model is the possibility that the company monitors, analyses, and interprets data on working conditions of human acts and physiological or ergonomic stress to workers (Armenteros-Cosme et al., 2025). This research supports evidence that 'raw' information is not potentially safe and can only become valuable when it is contextualized, processed in near real time, and returned to users in the form of interpretable indicators, alerts, or actions' suggestions (Fiegler-Rudol et al., 2025; Sánchez Vinces et al., 2025). Therefore, there are three closely coupled pieces – IoT, edge computing, and the on-the-fly processing of data in real time – that can be considered building the technological foundation for human-centred AI in industrial safety (Chen & Zheng, 2025; Nguyen et al., 2024).

In the field of occupational safety, the IoT includes numerous sensors and devices distributed throughout industrial work environments in order to monitor environmental, technological, or ergonomic parameters – i.e., noise level, temperature, vibration magnitude and direction, air quality values, as well as movement data (e.g., body posture) or selected physiological parameters (El-Helaly, 2024).

The literature points out that IoT greatly broadens an organization's 'perceptual horizon', capable of perceiving subtle changes and weak signals before humans can continuously recognize them (Abouhawwash et al., 2024; Rosak-Szyrocka et al., 2021; Shah & Mishra, 2024). Some of the documented applications include ergonomic load analysis, movement pattern analysis to reduce musculoskeletal disorders, and identifying slow thermal buildup in workplaces. Meanwhile, research utilities have made it clear that the efficiency of IoT is based on data processing infrastructure through which raw sensor streams are turned into human-readable and actionable information for people and organizations (Chen & Zheng, 2025; Kuchuk & Malokhvii, 2024).

Edge computing plays an imperative role in processing data not only on devices and nodes closer to where the data is generated but also at the cloud end systems (Arroba et al., 2024; Chen & Zheng, 2025). Such an architecture minimizes the latency associated with incident detection and warning delivery (e.g., potentially to as little as a few milliseconds), which is crucial in rapidly evolving scenarios such as machinery breakdown, structural instability, or sudden environmental change (Algarni et al., 2024; Pramanik et al., 2024). From a human-focused AI viewpoint, ultra-low latency is essential, as it enables information to be sent at the time when it would have its maximum protective value for the worker while reducing the likelihood of cognitive overload and decision delay (Nguyen et al., 2024).

IoT and edge computing in occupational safety systems. IoT and edge computing are widely used for integration into the occupational safety systems. In construction, it allows the workers' accurate localization to be achieved, tracking health parameters and warning them when they are within a dangerous zone (Chen & Zheng, 2025). For instance, in mining, IoT-edge architectures oversee the stability of rock-mass and environmental monitoring with the location of workers working on the hotspot (Armenteros-Cosme et al., 2025). Research also emphasizes the increasing relevance of wearables with integrated local analytics that offer the possibility of continuous monitoring of thermal stress, noise exposure, or non-ergonomic movement patterns, but at the same time, they reduce data transmission to external servers and increase privacy protections (Ghadi et al., 2025; Park & Kang, 2024).

The third pillar of the architecture, real-time data processing, allows local observations to be incorporated into a model of safety that is dynamic and continuously updated (Ahmed et al., 2025; Braniff et al., 2025; Chizubem et al., 2025). For predictive purposes, it facilitates the continuous updating of risk estimates by fusing historical information with real-time sensor readings. This capability transitions organizations from static risk assessment to dynamic as well as adaptive risk management (Mousavi et al., 2025; Yue et al., 2025). Real-time processing further improves event reconstruction and organizational learning, and therefore is a key component of next-generation safety-management ecosystems (Jetha et al., 2025). Table 2.3 presents a comparative analysis of IoT, edge computing, and real-time data processing within the context of human-centred industrial safety.

TABLE 2.3
A Comparative Analysis of IoT, Edge Computing, and Processing in Real Time within the Context of Human-Centred Industrial Safety

Element of Architecture	Role in Safety Systems	Relevance for Human-Centred AI	Examples
IoT	Gathering data about the work environment, ergonomics, human behaviours at work, health metrics, and machine states to broaden organizational awareness.	Detects weak signals at an early stage that may be overlooked by humans; helps companies make decisions and take preventive actions more accurately.	Measurement of noise, vibration, and thermal conditions; localization of workers on construction sites; detection of postural overload; atmosphere analysis in underground mines.
Edge Computing	Local data processing at source, lower latency; offload cloud facilities; immediate alerts.	Deliver warnings at the instant a human can respond, help in decision-making, avoid information overload, and respect privacy by processing locally.	Real-time notifications when entering dangerous areas and: Vibration analysis and early detection of machine breakdowns, wearables with on-device AI that calculate heat stress or noise exposure, or physiological overload.
Real-time data processing	Local processing near the source; low latency; offload cloud infrastructure; quick alerts.	Permits continued fine-tuning of protective choices; increases situational awareness; empowers workers to feel in control with real time, understandable risk signals.	Chemical industry safety systems; pre-accident event sequence-capturing systems; IoT stream data-based dynamic risk analysis.

Source: Own study.

The examples in Table 2.3 show that the merging of IoT, edge computing, and real-time data processing is not a vague technological goal but well documented, integrated function of deployed systems operating in major industries with which we work. To better depict how this architecture works in practice – and why it supports safety mechanisms informed by the tenets of human-centred AI – three examples frequently debated in the recent scientific literature are provided here. Each instance illustrates a different aspect of the IoT-edge-real-time technology stack with examples showing worker empowerment, risk amplification, and informed decision-making.

CASE STUDY 1 ARCHITECTURE: IOT-EDGE SYSTEM FOR WORKER LOCALIZATION AND HAZARD DETECTION

The IoT appears to be the future for large construction projects, given the growing use of advanced safety systems with IoT sensors fitted in helmets, vests, and other items of protective clothing. These sensors capture critical safety measurements: workers' exact position in the worksite, their activity path, exposure to weather conditions, and distance to high-risk areas like excavation zones or crane/handling corridors.

The sensor information is processed by the edge modules placed in mobile AMs interconnected on-site. Real time, on-the-edge analytics allow the system to discover hazardous events a few seconds in advance of their occurrence and issue personalized alerts. The worker is assigned an instruction, for example:

> **'You're coming into danger territory – stop and turn back the other way'**.

This technique obviously enhances the capacity of situational awareness and permits immediate preventive action. A key aspect is that the edge system does not substitute human judgement – adhering to human-centred AI principles, it supplies information that complements, but doesn't replace, the ability of the worker to judge its rules-of-the-road.

(Chen & Zheng, 2025; Kanan et al., 2018)

CASE STUDY 2 MINING: IOT-EDGE MONITORING OF TUNNELS STABILITY AND ENVIRONMENTAL PARAMETERS

In deep underground mining, the communication-on-the-go capability is a prerequisite to ensure the safety and efficiency of highly mobile systems in potentially limited control environments. Internet of Things (IoT) platforms with sensors standing in tunnels and shafts collect deformation, micro-vibration, air quality, pressure, temperature, and workers' location.

These streams are also sent off instantly, in milliseconds, to underground edge units, providing analysis without those cloud-only delays. Signals of impending collapse, sudden pressure increases that could be of decakiloton yield or higher (21), and degradation of the atmosphere that is harmful are detected instantly with an alert as:

'Immediate evacuation: critical pressure rise has been detected'.

In mining, where seconds can be the difference between controlled event and a catastrophe, edge processing is the key.

(Armenteros-Cosme et al., 2025)

CASE STUDY 3 WEARABLES AND ERGONOMICS IN INDUSTRY 5.0: REGION-WISE MOTION ANALYSIS AND HEAT LOAD

In logistics and light manufacturing sectors, wearable IoT devices, including smart vests, wristbands, and joint-based sensors, are used. They record posture, work pace, range of motion, micro-vibrations, body temperature, outside temperature, and heat load.

Spread even further and combined with edge processing is a full-on local processing of data, ultimately significantly lower network needs (like all the way gone low), leading to improved employee privacy and security since there is no need to send data up anywhere. It gives instant ergonomic feedback, for example:

'Please maintain posture – 6 s excessive forward bending detected'
'Heat-stress risk – leave the high-temperature zone'.

Consistent with human-centred AI, the technology is non-prescriptive, assisting rather than directing behaviours, which supports individual self-regulation by informing workers of their state so that they are able to make informed, on-the-job decisions. Collected evidence indicates that AI can improve employee well-being and agency and reduce safety-systems burden through early recognition of biomechanical (e.g., posture) and environmental exposures.

(Nguyen et al., 2024)

2.3 DIGITAL TWINS AND CASCADING SAFETY ARCHITECTURE

A digital twin is one of the most popular concepts regarding the development of digital industrial safety systems. It refers to the digital replica of any physical object, process or system that can be used to simulate, monitor, and optimize in real time. For risk management, digital twins represent test-beds for the testing and validation of hazard detection and prediction algorithms to be pursued against regulatory constraints, and technical standards (Chizubem et al., 2025; Francis et al., 2024; Masi et al., 2023).

In the age of AI, edge computing, and distributed analytics, digital twins become an essential part of cascading safety architectures – a multi-level decision-making framework for safety management that includes a hierarchical flow of data and control. Within this structure, local layers (such as sensors and PLCs) and supervisory layers (cloud platforms, AI models, and expert systems) help to continuously adjust protection strategies relative to the current state of operation (Ahmed et al., 2025; Braniff et al., 2025).

Digital twins provide an environment for testing and refining AI models with synthetic but realistic operational data. With this, a shift is possible: from AI processing information to an analysis level where decisions are made. From this perspective, it becomes feasible to evaluate the performance of safety systems in situations that cannot be reproduced in real life, particularly extreme overloads, failures, and rare high-risk behaviours. The virtual testing has the benefits of cost savings, a flexible nature, and a highly reduced operational risk (Francis et al., 2024; Khdoudi et al., 2024).

Recent literature more and more underlines that it's interesting to combine digital twins with RL methods and AD approaches (Lian et al., 2023; Lu et al., 2020; Moharam et al., 2025; Zhang et al., 2024). This combination provides the possibility of mechanisms that adapt and validate safety models continuously from design through operation to updates and modifications (Elía & Pagola, 2025; Inamdar et al., 2024). Digital twins that effectively work in hazardous industrial settings must fulfil three fundamental requirements:

- Reliability models should be able to run in changing environmental conditions and handle input data errors while their outputs remain consistent over time (Zhang et al., 2025a);
- Interpretability – safety systems, particularly in regulated industries, want algorithmic visibility and understandable decision-making (Chandola et al., 2009; Zhu et al., 2025);
- Real-time validation – constantly validate simulation results by comparison with actual events; models can be updated rapidly, and protective measures enhanced (Agnusdei et al., 2021; Constantinescu et al., 2020; Hua et al., 2022).

Digital twins are increasingly being used as central elements of cascading safety architectures – overall solutions where safety-relevant decisions are taken at the same time at multiple levels (Kabashkin, 2025; Wang et al., 2024). A local sensor could sense that a vibration level has exceeded a threshold, and yet an AI model in the cloud can predict the probability of failure and order the production line to stop. The cascading safety architecture is built on a three-layered data processing paradigm EDGE-FOG-CLOUD (Bonomi et al., 2012; Kumar et al., 2024; Mokhtari et al., 2021; Vu Khanh et al., 2023). Figure 2.2 presents a cascading safety architecture.

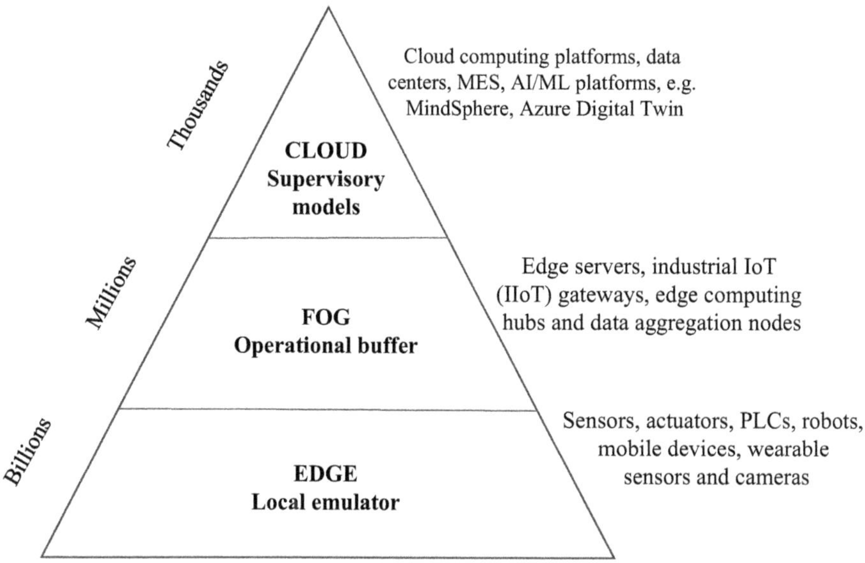

FIGURE 2.2 Cascading safety architecture.

(Own study based on Vu Khanh et al., 2023).

Each layer is responsible for corresponding roles of information processing, threat perception, and operational decision-making. Their combination gives rise to adaptive and failure-tolerant risk management systems in the industrial domain:

- Edge – provides the support to perform real-time processing right on devices (such as sensors and robots);
- Fog – aggregates data from a variety of local sources;
- Cloud – for long-term analysis and strategic planning with predictive analytics.

The edge is physical devices running at the point of action, within equipment, in production facilities, or on industrial machines. The data is processed on the platform, without using external servers, to reduce response time dramatically. Safety-wise, one can remark that this allows:

- immediate detection of abnormalities (e.g., high vibration, temperature peaks, human in risky zones),
- initiation of protection systems (such as an emergency machine shutdown),
- collaboration with prediction systems available on the local machine.

Typically, the application of a digital twin at the edge level could work as a local emulator that compares what is observed in real time with its expected behaviour and triggers an alert when there are critical deviations.

The FOG acts as an intermediate entity between local and cloud systems. It collects heterogeneous data from various sources and does pre-processing, cleaning, and unifying these multicategories' spatiotemporal correlation. Safety-specific functions include:

- and identifying risky behaviours across many spaces in parallel,
- co-ordinating protective measures between production units,
- filtering and labelling alarm events prior to sending them to the cloud.

FOG serves as an operational buffer between physical movement and strategic oversight, bearing the load for the cloud to provide quicker tactical decisions.

This multilayer architecture minimizes reaction delay and has more resilience within organizations and can adapt in terms of the protective actions according to different threat levels (Arroba et al., 2024; Park & Kang, 2024).

An example of how digital twins are applied in industrial safety systems is afforded by Siemens's offering built on top of the MindSphere platform, an Industrial IoT ecosystem that combines sensors, predictive models, and AI analytics. In a machine's double, for example, sensors feed data on how the machine is currently operating (mechanical vibration and temperature) into a digital model, which can then be compared to behaviour in reality – deviations here point to possible risks (Siemens, 2023). With the help of this architecture, the following benefits were possible:

- real-time anomaly discovery (e.g., micro-vibrations that precede mechanical failure),
- alarm call notification to operators when the safety limits are exceeded,
- preventive measures prior to the occurrences.

Reduced Order Model (ROM) based simulation approach allowed over 5,400x increase in processing speed with 200,000x less computing power demand, as reported by Siemens (2023). This algorithm accelerates predictive analytics, enabling it to go large scale for real-time execution within production systems.

As a result, this approach contributed to minimizing unplanned downtime, enhancing the predictability of operational decisions, and maximizing risk management at Siemens' European plants. This example has demonstrated the utility of digital twins as model-based predictive testing platforms and adaptive safety analytics (Chaparro-Cárdenas et al., 2025; Hossain et al., 2025; Mikołajewska et al., 2025; Rojas et al., 2025).

REFERENCES

Abouhawwash, M., Rosak-Szyrocka, J., & Gupta, S. K. (2024). *Aspects of Quality Management in Value Creating in the Industry 5.0 Way*: CRC Press.

Agnusdei, G. P., Elia, V., & Gnoni, M. G. (2021). A classification proposal of digital twin applications in the safety domain. *Computers & Industrial Engineering*, *154*, 107137.

Ahmed, M. D. M., Almansoori, S. H. M., Raza, A., & Badidi, E. (2025). Real-time pedestrian safety enhancement using edge AI in smart urban environments. In *2025 International Conference on Activity and Behavior Computing (ABC)* (pp. 1–8). IEEE.

Algarni, A., Acarer, T., & Ahmad, Z. (2024). An edge computing-based preventive framework with machine learning- integration for anomaly detection and risk management in maritime wireless communications. *IEEE Access*, *12*, 53646–53663.

Armenteros-Cosme, P., Arias-González, M., Alonso-Rollán, S., Márquez-Sánchez, S., & Carrera, A. (2025). Advancements in artificial intelligence and machine learning for occupational risk prevention: A systematic review on predictive risk modeling and prevention strategies. *Sensors*, *25*(17).

Arroba, P., Buyya, R., Cárdenas, R., Risco-Martín, J. L., & Moya, J. M. (2024). Sustainable edge computing: Challenges and future directions. *Software: Practice and Experience*, *54*(11), 2272–2296.

Bonomi, F., Milito, R., Zhu, J., & Addepalli, S. (2012). Fog computing and its role in the internet of things. In *Proceedings of the First Edition of the MCC Workshop on Mobile Cloud Computing* (pp. 13–16). New York: ACM.

Braniff, A., Akundi, S. S., Liu, Y., Dantas, B., Niknezhad, S. S., Khan, F., et al. (2025). Real-time process safety and systems decision-making toward safe and smart chemical manufacturing. *Digital Chemical Engineering*, *15*, 100227.

Ch, R., Sudheer, P., Batra, I., & Sembiring, F. (2025). A novel adaptive cluster-based federated learning framework for anomaly detection in VANETs. *Engineering Proceedings*, *107*(1), 79. https://www.mdpi.com/2673-4591/107/1/79

Chandola, V., Banerjee, A., & Kumar, V. (2009). Anomaly detection. *ACM Computing Surveys*, *41*(3), 1–58.

Chaparro-Cárdenas, S. L., Ramirez-Bautista, J.-A., Terven, J., Córdova-Esparza, D.-M., Romero-Gonzalez, J.-A., Ramírez-Pedraza, A., & Chavez-Urbiola, E. A. (2025). A technological review of digital twins and artificial intelligence for personalized and predictive healthcare. *Healthcare*, *13*(14), 1763. https://www.mdpi.com/2227-9032/13/14/1763

Chen, G., & Zheng, J. (2025). Real-time safety monitoring system for smart construction sites based on edge computing and multi-source sensing. In *Proceedings of the 2025 International Conference on Artificial Intelligence and Smart Manufacturing* (pp. 776–781). New York: ACM.

Cherry, K. M., & Qian, L. (2025). Supervised learning in DNA neural networks. *Nature*, *645*(8081), 639–647. https://www.nature.com/articles/s41586-025-09479-w

Chizubem, B., Subbiah, A., Izuchukwu, O. C., & Musa, K. S. (2025). Real-time monitoring using digital platforms for enhanced safety in hydrogen facilities – Current perspectives and future directions. *International Journal of Hydrogen Energy*, *98*, 487–499.

Coelho, D., Costa, D., Rocha, E. M., Almeida, D., & Santos, J. P. (2022). Predictive maintenance on sensorized stamping presses by time series segmentation, anomaly detection, and classification algorithms. *Procedia Computer Science*, *200*, 1184–1193.

Constantinescu, C., Giosan, S., Matei, R., & Wohlfeld, D. (2020). A holistic methodology for development of real-time digital twins. *Procedia CIRP*, *88*, 163–166.

Da Silva, M. P., Correa, S. P., Schaefer, M. A., Reis, J. C., Nunes, I. M., dos Santos, J. A., & Oliveira, H. N. (2025). Advancing agricultural remote sensing: A comprehensive review of deep supervised and self-supervised learning for crop monitoring. *Computers & Graphics*, *133*, 104434.

Dong, Y., Zhang, L., Farah, H., Zgonnikov, A., & van Arem, B. (2025). Data-driven semi-supervised machine learning with safety indicators for abnormal driving behavior detection. *Transportation Research Record: Journal of the Transportation Research Board*, *2679*(5), 319–334.

Dutta, D., & Mishra, S. K. (2024). Bots for mental health: The boundaries of human and technology agencies for enabling mental well-being within organizations. *Personnel Review*, *53*(5), 1129–1156.

El-Helaly, M. (2024). Artificial intelligence and occupational health and safety, benefits and drawbacks. *La Medicina del lavoro*, *115*(2), e2024014.

Elía, I., & Pagola, M. (2025). Anomaly detection in smart-manufacturing era: A review. *Engineering Applications of Artificial Intelligence*, *139*, 109578.

Fiegler-Rudol, J., Lau, K., Mroczek, A., & Kasperczyk, J. (2025). Exploring human-AI dynamics in enhancing workplace health and safety: A narrative review. *International Journal of Environmental Research and Public Health*, *22*(2), 199. https://www.mdpi.com/1660-4601/22/2/199

Fisher, E., Flynn, M. A., Pratap, P., & Vietas, J. A. (2023). Occupational safety and health equity impacts of artificial intelligence: A scoping review. *International Journal of Environmental Research and Public Health*, *20*(13), 6221. https://www.mdpi.com/1660-4601/20/13/6221

Francis, D., Friederich, J., Uhrmacher, A., & Lazarova-Molnar, S. (2024). Reinforcement learning for digital twins. *Digital Twins, Simulation, and the Metaverse*, 51–68. https://doi.org/10.1007/978-3-031-69107-2_3

Gan, H., Guo, L., Xia, S., & Wang, T. (2020). A hybrid safe semi-supervised learning method. *Expert Systems with Applications*, *149*, 113295.

Gan, H., Li, Z., Wu, W., Luo, Z., & Huang, R. (2018). Safety-aware graph-based semi-supervised learning. *Expert Systems with Applications*, *107*, 243–254.

Gan, H., Yang, Z., & Zhou, R. (2023). Adaptive safety-aware semi-supervised clustering. *Expert Systems with Applications*, *212*, 118751.

Ghadi, Y. Y., Shah, S. F. A., Waheed, W., Mazhar, T., Ahmad, W., Saeed, M. M., & Hamam, H. (2025). Integration of wearable technology and artificial intelligence in digital health for remote patient care. *Journal of Cloud Computing*, *14*(1), 39. https://doi.org/10.1186/s13677-025-00759-4

Grunova, D., Bakratsi, V., Vrochidou, E., & Papakostas, G. A. (2024). Machine learning for anomaly detection in industrial environments. *Engineering Proceedings*, *70*(1), 25. https://www.mdpi.com/2673-4591/70/1/25

Guan, D., Xing, Y., Huang, J., Xiao, A., El Saddik, A., & Lu, S. (2025). S2Match: Self-paced sampling for data-limited semi-supervised learning. *Pattern Recognition*, *159*, 111121.

Guo, K., Yang, Z., Yu, C.-H., & Buehler, M. J. (2021). Artificial intelligence and machine learning in design of mechanical materials. *Materials horizons*, *8*(4), 1153–1172.

Hossain, M. A., Hasan, T., Karovic, V., Abdeljaber, H. A. M., Haque, M. A., Ahmad, S., et al. (2025). Deep learning and ensemble methods for anomaly detection in ICS security. *International Journal of Information Technology*, *17*(3), 1761–1775.

Hua, E. Y., Lazarova-Molnar, S., & Francis, D. P. (2022). Validation of digital twins: Challenges and opportunities. In *2022 Winter Simulation Conference (WSC)* (pp. 2900–2911). IEEE.

Inamdar, R., Sundarr, S. K., Khandelwal, D., Sahu, V. D., & Katal, N. (2024). A comprehensive review on safe reinforcement learning for autonomous vehicle control in dynamic environments. *e-Prime - Advances in Electrical Engineering, Electronics and Energy*, *10*, 100810.

Jetha, A., Bakhtari, H., Irvin, E., Biswas, A., Smith, M. J., Mustard, C., et al. (2025). Do occupational health and safety tools that utilize artificial intelligence have a measurable impact on worker injury or illness? Findings from a systematic review. *Systematic Reviews*, *14*(1), 146. https://doi.org/10.1186/s13643-025-02869-1

Kabashkin, I. (2025). Mathematical framework for digital risk twins in safety-critical systems. *Mathematics*, *13*(19), 3222. https://www.mdpi.com/2227-7390/13/19/3222

Kamat, P., & Sugandhi, R. (2020). Anomaly detection for predictive maintenance in Industry 4.0- A survey. In *E3S Web of Conferences* (p. 170).

Kammerer, K., Hoppenstedt, B., Pryss, R., Stökler, S., Allgaier, J., & Reichert, M. (2019). Anomaly detections for manufacturing systems based on sensor data-insights into two challenging real-world production settings. *Sensors*, *19*(24), 5370. https://www.mdpi.com/1424-8220/19/24/5370

Kanan, R., Elhassan, O., & Bensalem, R. (2018). An IoT-based autonomous system for workers' safety in construction sites with real-time alarming, monitoring, and positioning strategies. *Automation in Construction*, *88*, 73–86.

Khdoudi, A., Masrour, T., El Hassani, I., & El Mazgualdi, C. (2024). A deep-reinforcement-learning-based digital twin for manufacturing process optimization. *Systems*, *12*(2), 38. https://www.mdpi.com/2079-8954/12/2/38

Kuchuk, H., & Malokhvii, E. (2024). Integration of IOT with Cloud, fog, and edge computing: A Review. *Advanced Information Systems*, *8*(2), 65–78. http://ais.khpi.edu.ua/article/view/305471

Kumar, N., Pal, S. K., Agarwal, P., Rosak-Szyrocka, J., & Jain, V. (2024). *Human-Machine Collaboration and Emotional Intelligence in Industry 5.0*. IGI Global.

Li, J., Izakian, H., Pedrycz, W., & Jamal, I. (2021). Clustering-based anomaly detection in multivariate time series data. *Applied Soft Computing*, *100*, 106919.

Lian, Y., Geng, Y., & Tian, T. (2023). Anomaly detection method for multivariate time series data of oil and gas stations based on digital twin and MTAD-GAN. *Applied Sciences*, *13*(3), 1891. https://www.mdpi.com/2076-3417/13/3/1891

Lu, Q., Xie, X., Parlikad, A. K., & Schooling, J. M. (2020). Digital twin-enabled anomaly detection for built asset monitoring in operation and maintenance. *Automation in Construction*, *118*, 103277.

Masi, M., Sellitto, G. P., Aranha, H., & Pavleska, T. (2023). Securing critical infrastructures with a cybersecurity digital twin. *Software and Systems Modeling*, *22*(2), 689–707.

Mikołajewska, E., Mikołajewski, D., Mikołajczyk, T., & Paczkowski, T. (2025). Generative AI in AI-based digital twins for fault diagnosis for predictive maintenance in Industry 4.0/5.0. *Applied Sciences*, *15*(6), 3166. https://www.mdpi.com/2076-3417/15/6/3166

Mitchell, T. M. (2010). *Machine Learning* (International ed., [Reprint.]). McGraw-Hill Series in Computer Science. New York: McGraw-Hill.

Moharam, M. H., Hany, O., Hany, A., Mahmoud, A., Mohamed, M., & Saeed, S. (2025). Anomaly detection using machine learning and adopted digital twin concepts in radio environments. *Scientific Reports*, *15*(1), 18352. https://www.nature.com/articles/s41598-025-02759-5

Mokhtari, S., Abbaspour, A., Yen, K. K., & Sargolzaei, A. (2021). A Machine learning approach for anomaly detection in industrial control systems based on measurement data. *Electronics*, *10*(4), 407. https://www.mdpi.com/2079-9292/10/4/407

Montealegre-López, N. (2025). Exploring the role of trust in AI-driven decision-making: A systematic literature review. *Management Review Quarterly*, 1–51. https://doi.org/10.1007/s11301-025-00526-4

Mousavi, M., Shen, X., Zhang, Z., Barati, K., & Li, B. (2025). IoT-Bayes fusion: Advancing real-time environmental safety risk monitoring in underground mining and construction. *Reliability Engineering & System Safety*, *256*, 110760.

Nassif, A. B., Talib, M. A., Nasir, Q., & Dakalbab, F. M. (2021). Machine learning for anomaly detection: A systematic review. *IEEE Access*, *9*, 78658–78700.

Neupane, D., Bouadjenek, M. R., Dazeley, R., & Aryal, S. (2024). A comparative study of semi-supervised anomaly detection methods for machine fault detection. In *PHM Society European Conference* (Vol. *8*, No. 1), p. 10).

Nguyen, T., Nguyen, D. H., Nguyen, Q.-T., Tran, K. D., & Tran, K. P. (2024). Human-centered edge AI and wearable technology for Workplace health and safety in Industry 5.0. In K. P. Tran (Ed.), *Springer Series in Reliability Engineering. Artificial Intelligence for Safety and Reliability Engineering* (pp. 171–183). Cham: Springer Nature Switzerland.

Orlov, A. A., Akhmetshin, T. N., Horvath, D., Marcou, G., & Varnek, A. (2025). From high dimensions to human insight: Exploring dimensionality reduction for chemical space visualization. *Molecular Informatics*, *44*(1), e202400265.

Park, J., & Kang, D. (2024). Artificial intelligence and smart technologies in safety management: A comprehensive analysis across multiple industries. *Applied Sciences*, *14*(24), 11934. https://www.mdpi.com/2076-3417/14/24/11934

Peng, D., Desmet, W., & Gryllias, K. (2025). Reconstruction-based deep unsupervised adaptive threshold support vector data description for wind turbine anomaly detection. *Reliability Engineering & System Safety*, *260*, 110995.

Pramanik, P. K. D., Pal, S., Mukhopadhyay, M., & Choudhury, P. (2024). Sustainable edge computing with mobile crowd computing: A proof-of-concept with a smart HVAC use case. *The Journal of Supercomputing*, *80*(16), 23911–23994.

Pu, G., Wang, L., Shen, J., & Dong, F. (2021). A hybrid unsupervised clustering-based anomaly detection method. *Tsinghua Science and Technology*, *26*(2), 146–153.

Qian, C., Tang, W., & Wang, Y. (2025). RGAnomaly: Data reconstruction-based generative adversarial networks for multivariate time series anomaly detection in the Internet of Things. *Future Generation Computer Systems*, *167*, 107751.

Quatrini, E., Costantino, F., Di Gravio, G., & Patriarca, R. (2020). Machine learning for anomaly detection and process phase classification to improve safety and maintenance activities. *Journal of Manufacturing Systems*, *56*, 117–132.

Raptis, T. P., Passarella, A., & Conti, M. (2019). Data management in Industry 4.0: State of the art and open challenges. *IEEE Access*, *7*, 97052–97093.

Razzaq, K., & Shah, M. (2025). Machine learning and deep learning paradigms: From techniques to practical applications and research frontiers. *Computers*, *14*(3), 93. https://www.mdpi.com/2073-431x/14/3/93

Reinforcement learning: An introduction (1998).

Rojas, L., Peña, Á., & Garcia, J. (2025). AI-driven predictive maintenance in mining: A systematic literature review on fault detection, digital twins, and intelligent asset management. *Applied Sciences*, *15*(6), 3337. https://www.mdpi.com/2076-3417/15/6/3337

Rosak-Szyrocka, J., Zywiolek, J., Kulinska, E., & Matulewski, M. (2021). Analysis of enterprises' readiness in for Industry 4.0 implementation: The case of Poland. *European Research Studies Journal*, *XXIV*(3), 615–628. https://www.um.edu.mt/library/oar/handle/123456789/83208

Sánchez Vinces, B. V., Schubert, E., Zimek, A., & Cordeiro, R. L. F. (2025). A comparative evaluation of clustering-based outlier detection. *Data Mining and Knowledge Discovery*, *39*(2), 13. https://doi.org/10.1007/s10618-024-01086-z

Serani, A., & Diez, M. (2025). A survey on design-space dimensionality reduction methods for shape optimization. *Archives of Computational Methods in Engineering*, 1–28. https://doi.org/10.1007/s11831-025-10349-x

Serradilla, O., Zugasti, E., Ramirez de Okariz, J., Rodriguez, J., & Zurutuza, U. (2021). Adaptable and explainable predictive maintenance: Semi-supervised deep learning for anomaly detection and diagnosis in press machine data. *Applied Sciences*, *11*(16), 7376. https://www.mdpi.com/2076-3417/11/16/7376

Shah, I. A., & Mishra, S. (2024). Artificial intelligence in advancing occupational health and safety: An encapsulation of developments. *Journal of Occupational Health*, *66*(1).

Siemens AG. (2023). *AI and the Digital Twin: Turbocharging the Digital Enterprise*. https://www.siemens.com/industrial-ai

Su, T., Wu, T., Zhao, J., Scaglione, A., & Le Xie (2025). A review of safe reinforcement learning methods for modern power systems. *Proceedings of the IEEE*, *113*(3), 213–255.

Sun, X., Li, R., & Yuan, Z. (2020). Anomaly detection in discrete manufacturing systems by pattern relation table approaches. *Sensors*, *20*(20), 5766. https://www.mdpi.com/1424-8220/20/20/5766

Tong, C., Wang, Y., Liu, H., Tong, B., Zhu, J., & Kou, M. (2025). From risk management to AI-driven safety: A bibliometric review of occupational safety and health research. *Results in Engineering*, *28*, 107440.

Vivian, G. A., Bauder, R. A., & Khoshgoftaar, T. M. (2025). A comprehensive survey on machine learning for workplace injury analysis: Risk prediction, return to work strategies, and demographic insights. *Journal of Big Data*, *12*(1), 167. https://doi.org/10.1186/s40537-025-01229-z

Vu Khanh, Q., Vi Hoai, N., Dang Van, A., & Nguyen Minh, Q. (2023). An integrating computing framework based on edge-fog-cloud for internet of healthcare things applications. *Internet of Things*, *23*, 100907.

Wabersich, K. P., & Zeilinger, M. N. (2021). A predictive safety filter for learning-based control of constrained nonlinear dynamical systems. *Automatica*, *129*, 109597.

Wang, R., Venugopalan, S., & Adepu, S. (2024). Safety analysis for cyber-physical systems under cyber attacks using digital twin. In *2024 IEEE International Conference on Cyber Security and Resilience (CSR)* (pp. 1–8). IEEE.

Wani, A. A. (2025). Comprehensive review of dimensionality reduction algorithms: Challenges, limitations, and innovative solutions. *PeerJ Computer Science*, *11*, e3025. https://peerj.com/articles/cs-3025/

Wei, Z., Liu, H., Tao, X., Pan, K., Huang, R., Ji, W., & Wang, J. (2023). Insights into the application of machine learning in industrial risk assessment: A bibliometric mapping analysis. *Sustainability*, *15*(8), 6965. https://www.mdpi.com/2071-1050/15/8/6965

Yuan, S. (2025). Research on anomaly detection and privacy protection of network security data based on machine learning. *Procedia Computer Science*, *261*, 227–236.

Yue, X., Liao, Y., Peng, H., Kang, L., & Zeng, Y. (2025). A high-dimensional feature selection algorithm via fast dimensionality reduction and multi-objective differential evolution. *Swarm and Evolutionary Computation*, *94*, 101899.

Zhang, H., Zhang, R., & Sun, J. (2025a). Developing real-time IoT-based public safety alert and emergency response systems. *Scientific Reports*, *15*(1), 29056. https://www.nature.com/articles/s41598-025-13465-7

Zhang, J. (2025). A survey of semi-supervised and unsupervised learning methods for industrial defect anomaly detection. *Journal of Computer Science and Frontier Technologies*, *1*(1), 1.

Zhang, J., Zhang, W., Liao, W., Zhao, T., & Miao, Y. (2025b). Is the real-time data of process safety reliable? An anomaly detection method based on the graph neural network. *Process Safety and Environmental Protection*, *198*, 107066.

Zhang, L., Yang, C., Yan, Y., Cai, Z., & Hu, Y. (2024). Automated guided vehicle dispatching and routing integration via digital twin with deep reinforcement learning. *Journal of Manufacturing Systems*, *72*, 492–503.

Zhu, Q., Ding, Y., Jiang, J., & Yang, S.-H. (2025). Anomaly detection using invariant rules in industrial control systems. *Control Engineering Practice*, *154*, 106164.

3 Intelligent Hazard Sensing and Environmental Screening

3.1 INTELLIGENT SENSOR NETWORKS AND DATA FUSION

Among many, ISNs represent one of the most revolutionary developments in the metamorphosis of the architecture of industrial safety systems. A distributed network of sensing nodes integrated with computational intelligence is able to realize, understand, and predict hazardous conditions much earlier than they could turn into an accident or operational failure. Contrasting traditional measurement systems operating in silos and requiring interpretation by human beings, ISNs were designed as interconnected ecosystems in which sensors collaborate through machine learning algorithms, wireless communication, and dynamic data fusion. The purpose of such networks is not just the collection of environmental or mechanical data but to convert it into actionable knowledge that will support immediate decisions at the operational level. In this sense, ISNs constitute the perceptual foundation of this emerging paradigm now known as Intelligent Safety Management (ISM), integrating cyber-physical awareness with real-time risk prediction (Ding et al., 2025; Li et al., 2025a; Zhang et al., 2020).

It follows that the conceptual development of intelligent sensing in industrial systems occurs in three steps: from reactive sensing, via proactive sensing, to today's cognitive sensing. The first could make single-parameter measurements – pressure, vibration, or temperature – whose outputs were elaborated by either a human or SCADA systems in a delayed analytical cycle. A new stage had introduced device connectivity and elementary analytics, whereby sensors started comparing themselves with each other on a peer-to-peer basis, reporting threshold deviation alerts. The present stage is that of cognitive sensing, where sensor networks can reason and learn from data: sensors act, in virtue of artificial intelligence and data fusion mechanisms, as constituent components of a distributed cognitive entity capable of performing an adaptive calibration, contextual interpretation, and predictive forecasting. In other words, this is a change in the very role played by sensors: no longer passive measuring devices but rather intelligent agents who take part in some collective situational awareness.

The general architecture of intelligent sensor networks from the system viewpoint is hierarchical. At the lowest level, primary sensors operate, monitoring physical variables such as temperature, gas concentration, humidity, vibration amplitude, or noise intensity. Above them is the edge processing layer; this is where microcontrollers or embedded AI modules pre-process the signals, clean noise, operate on statistical filters, and perform local classification algorithms. Such local processing

 DOI: 10.1201/9781003754213-4

will dramatically reduce latency in data and thus enable immediate actuation based on abnormal events, like thermal spikes or toxic gas accumulation, without waiting for central approval. Further, the layer of data aggregation and fusion hubs is where heterogeneous signals merge through the use of some mathematical models, such as Bayesian inference, Kalman filtering, or deep learning-based feature fusion. The top layer is the decision intelligence unit that transforms fused data into high-level insights, visual dashboards, and predictive indicators to be passed on to the control systems, such as SCADA or DCS.

The process that is very different in an intelligent network, compared with more conventional monitoring architectures, is the process of data fusion. It is defined here as a process of merging multi-variate data streams into one analytic representation, including environmental, mechanical, and physiological. For example, a sudden increase in machine vibration, with a slight increase in temperature, coupled with a corresponding increase in local CO_2, might indicate early-stage bearing degradation or frictional overheating well before any kind of catastrophic failure. Each of these in isolation may not result in an alarm; passing through the fusion model, however, produces latent hazard patterns. Data fusion happens at three levels: low-level fusion, where raw sensor data are combined; feature-level fusion, which integrates extracted statistical or spectral features; and decision-level fusion, which integrates independent classifier outcomes. There has been an increased usage of neural network architectures, notably convolutional neural networks (CNNs) and long short-term memory networks (LSTMs), hence increasing the nonlinear relationships that can be detected in systems and temporal dependencies over many sensor modalities (Kong & Li 2025; Tabrizchi & Kuchaki Rafsanjani, 2025; Wei & Yang 2025).

The advantages of intelligent sensor networks are that sensor readings can be contextualized. Most traditional systems are subject to false alarms because they apply fixed thresholds, considering no external context. On the contrary, AI-driven ISNs analyse conditions holistically, recognizing that a rise in temperature can be normal at certain phases in production or highly vibrating conditions can be acceptable under certain loads. The advantage of adaptability over time is that acceptable ranges dynamically readjust themselves via machine learning algorithms based on experience accumulated and continuously optimize such functions. Such adaptive intelligence somewhat resembles the process of human perception, whereby cognition refines sensory interpretation with respect to context, memory, and expectations (Angel Joseph & Sivaraju, 2025; Chen et al., 2025; Cui et al., 2025; Wang & Huang, 2025).

But all the time, it gives another dimension to the intelligence of the network: biological and human-related sensing. Sensors integrated into wearables or personal protective equipment monitor workers' physiological parameters: heart rate, skin temperature, eye movements, or signs of fatigue via micro-vibrations and motion patterns. Complemented by environmental data, ambient temperature, or air quality, the system can predict states of physical overload, dehydration, or oxygen deficiency. All these integrations of multimodal capabilities create a cyber-physical shield around each worker in high-risk sectors like mining, metallurgy, or chemical processing.

The biological and mechanical domains unite for holistic safety in machines and human co-actors in the same ecosystem. They rely, from the technological point of view, on four aspects: sensor accuracy, communication latency, network resilience, and data interpretability. The accuracy is maintained by calibration algorithms that detect drift or interference and adjust correspondingly. Due to edge computing, communication latency remains minimal, since initial analytics are performed locally rather than in faraway data centres. Network resilience, assured through redundancy and mesh topologies, permits operation in the event of failure of individual nodes. These issues will be relevant to humans sooner or later: interpretability means finding out why the system issued a certain alert. The XAI method shall be used wherever possible, since this is crucial for tracing decision pathways from sensor readings right up to the logic at the generation of an alert. This builds human trust and opens up doors for greater regulatory acceptance under the ISO 45001 and IEC 61508 frameworks concerning functional safety (Liu et al., 2025a; Wang et al., 2025b; Xi et al., 2025).

The industrial deployment of intelligent sensor networks has redefined the paradigm of maintenance. From periodic inspections, today's systems operate on condition-based or predictive maintenance regimes where interventions are initiated based on real-time degradation patterns and not on arbitrary time intervals. This cuts costs by avoiding unscheduled downtimes and prolonging the life of critical assets. Data fusion from vibration sensors, infrared thermography, acoustic emissions, and lubricant quality indicators enables high-fidelity prognostics – that is, predicting failure probability with temporal precision. Integration with SCADA or enterprise maintenance systems ensures that such predictions are not isolated insights but operational commands, thus automatically triggering maintenance tickets or adjusting process parameters.

A representative example of such architecture is the AI-driven sensing network installed by Duke Energy in the US. It has integrated a multi-tier system that connects turbine, transformer, and substation sensors through an edge AI infrastructure performing vibration spectra and thermal pattern analyses locally. Thus, machine learning models continuously compare the current signature with historical baselines to estimate deviations long in advance of mechanical failure. When notified of an anomaly, a notice is then sent to the central control room, where the scheduling of maintenance priorities has been automatically updated. The result has been a quantifiable decrease in unplanned outages, accompanied by a remarkable increase in operational efficiency. More importantly, the approach undertaken by Duke Energy has succeeded in demonstrating that intelligent sensing is not confined to single facilities but can be effectively scaled across entire transmission and generation networks, thus making industrial safety a data-driven and predictive enterprise.

Another example of AI-based safety analytics has been initiated by Anglo American Mining in some of its platinum mines in South Africa: Intelligent sensor arrays track gaseous emissions, vibration, and the concentration of micro-dust in underground tunnels. The system recognizes precursors of hazardous conditions – for example, a sudden rise in methane and particulate matter with reduced airflow velocity. The alerts get sent to local devices, which provide situational awareness right away to workers and supervisors. With time, the network learns the particular spatial and temporal patterns of the mine environment and can thus distinguish between normal fluctuations and genuine hazards. This is only one example of where machine perception and environmental adaptation meet within the salient feature of Industry 4.0's safety paradigm reached in the works of Ding et al. (2025), Macrae (2025), Wolniak et al. (2024b), and Wolniak et al. (2025).

The development of intelligent sensor networks also raises several challenges, above all methodological. First, there is scalability, meaning the ability to maintain precision and low latency when thousands of sensors may be operating in dynamic industrial contexts. Second, semantic interoperability means different sensors, protocols for communication, and AI models need to be able to converse with no loss of meaning. Third, there is cybersecurity – for every additional sensor connected, the attack surface grows greater, and anomaly detection via AI will have to grow beyond mechanical and environmental data to network integrity. Finally, ethics and privacy issues arise when biological or behavioural data relating to workers forms part of safety analytics. In this regard, transparent governance, anonymization, and adherence to occupational data protection standards will be as important as technical performance.

Table 3.1 shows the relationship between ISNs and data fusion frameworks to organize both the organizational and operational dimensions regarding the management of industrial safety. They represent layers of system intelligence to be implemented, including physical sensing and communication to contextual reasoning and

TABLE 3.1
Features and Functional Dimensions of Intelligent Sensor Networks and Data Fusion

Dimension/Feature	Description	Industrial Significance and Application
Network topology	Distributed, hierarchical organization of sensors with multi-hop and redundant communications.	Ensures continuous operation in unfavourable conditions and resilience against node failures.
Sensor heterogeneity	Integration of different sensor modalities: environmental, mechanical, chemical, and biological.	Allow monitoring of workplace conditions and system performance from many dimensions.
Edge intelligence	Local data processing with AI algorithms embedded in sensors or gateways.	Reduces latency, decreases network load, and enables real-time hazard detection.
Data fusion mechanisms	Combination of multi-source data using Bayesian inference, Kalman filters, or deep learning.	Provides holistic situational awareness with early recognition of anomalies.
Context awareness	Dynamic adaptation of thresholds and interpretations depending on the operational context.	This reduces false alarms and improves accuracy in complicated variable conditions.
Self-calibration	This involves autonomous adjustment of sensors based on feedback or comparative analysis.	Ensures the accuracy of measurement and limits the need for maintenance.
Communication protocols	Use of industrial standards like OPC UA, MQTT, or Modbus TCP/IP.	Ensures interoperability with SCADA and Industrial Internet of Things platforms.
Energy efficiency	Energy consumption optimization through duty-cycling and computation at the local node:	Extends the working lifetime of wireless sensors, especially in remote areas.

(Continued)

TABLE 3.1 (CONTINUED)

Dimension/Feature	Description	Industrial Significance and Application
Security and integrity	Implementations of encryption, authentication, and anomaly-based intrusion detection.	Protects sensor data and network integrity against cyber threats.
Scalability	Ability to incorporate thousands of nodes without degradation in performance.	Allows installations on large industrial plants or in dispersed energy systems.
Fault tolerance	Redundant nodes and adaptive routing mechanisms to sustain functionality under failure.	Prevents data loss and maintains continuity in the monitoring of safety.
Fusion levels	Multi-tier approach: raw-data fusion, feature-level fusion, and decision-level fusion.	Enables complex reasoning from low-level measurements to high-level conclusions.
Predictive modelling	Application of machine learning to predict failures or abnormal behaviours.	It forms the analytical basis for predictive maintenance and risk prevention.
Human-machine interaction	Provide visualizations of fused data: dashboards, AR interfaces, or digital twins.	It intuitively allows the operator to make decisions based on the interpretation of risk.
Standard compliance	Conforms to ISO 45001 (safety), IEC 61508 (functional safety), and ISO 13374 (condition monitoring).	Ensures compliance with regulations and integration into established industrial frameworks.

Source: Author's own work.

predictive control. Their integration into one cohesive cyber-physical ecosystem enables real-time perception and interpretation of environmental and mechanical signals. Thus, adaptive, self-learning, and resilient monitoring infrastructures will become the backbone of state-of-the-art Industry 4.0 and 5.0 safety paradigms.

3.2 EDGE-BASED ANALYTICS AND LATENCY-FREE ALERTS

The fundamental principle of edge computing is a strategic shift from centralized data analysis to decentralized, proximity-based computation. From this perspective, edge computing is increasingly the basis for the evolution of intelligent safety systems in industrial contexts. Edge analytics means processing sensor data directly at or near the location where it was generated – that is, from the machinery, control units, or sensor nodes themselves. Such an architectural shift helped solve one of the main weaknesses of cloud-centric approaches to safety: latency. A latency of even several seconds between capture and execution of relevant decisions can make all the difference between taking anticipatory action or experiencing a calamity in hazardous industrial contexts. This is avoided by the adoption of edge analytics, whereby real-time detection, interpretation, and reaction are enabled to build latency-free alerts.

Traditional industrial monitoring consists of the field sensors sending a stream of continuous data to central servers or data centres, where machine learning algorithms run to detect anomalies and predict failures. Such a model is useful for long-term analytics and global optimization but represents major communications bottlenecks, bandwidth constraints, and response delays. Conversely, edge analytics embeds computational intelligence into local devices that contain integrated microprocessors and AI accelerators, able to execute advanced signal processing, pattern recognition, and threshold-based decision-making autonomously from the cloud infrastructure. Only necessary or aggregated information is transmitted upwards to supervisory systems, reducing the load on the network vastly, at the same time as ensuring that operational safety will be maintained under conditions of limited connectivity.

Architecture for edge-based analytics in safety systems can be characterized by three layers: edge, fog, and cloud. The edge layer contains analytics embedded directly into sensors, smart controllers, or industrial gateways. The fog layer acts as an intermediate aggregation point processing data from several edge nodes for regional or process-specific optimization. The cloud would remain responsible for high-level coordination, historical analysis, and policy updates. Intelligence will relate to a system that is able to determine autonomously how much data needs to be processed locally or how much needs to be aggregated and stored long term in order to provide global correlation. Further, this will then be the hallmark of distributed cognition, whereby each component of the network contributes to situational awareness (Hichri et al., 2025; Jonek-Kowalska & Wolniak, 2021; Rosak-Szyrocka & Wolniak, 2025b; Ye et al., 2025b).

The key functions of edge analytics in industrial safety systems will be signal filtering, pattern classification, anomaly detection, and local actuation. First is signal filtering, which cleans the raw-data noise and removes the irrelevant fluctuations that can be caused by mechanical vibration or electromagnetic interference. The second function will be to make sense of the preprocessed signal with local AI models – for example, discrimination between normal operational vibration and early signs of bearing wear. Third, anomaly detection kicks in when the readings start to deviate from established patterns. Subtle abnormalities not so obvious to people get picked up in respect of this, with the help of algorithms based on unsupervised learning by drawing upon autoencoders or clustering techniques. Finally, independent local actuation by the edge node may trigger an alarm, initiate an emergency stop, or even run a parameter adjustment without waiting for any kind of instruction from the central control room. Such immediate actions turn safety systems from passive observers to proactive agents (El-Nabi et al., 2025; Hasrat et al., 2026; Jonek-Kowalska & Wolniak, 2025; Ye et al., 2025a).

Where reaction times spell the difference in worker safety and equipment integrity, latency-free notification is of prime importance in high-risk sectors. Analytics at the edge reduces latency from seconds down to milliseconds, thus achieving near-instant response times. Consider a gas sensor with a locally attached edge processor at an oil refinery or chemical plant that detects a dangerous spike in volatile organic compounds: it can turn ventilation systems on or trigger alarms before the information has arrived in the control centre. On turbines in power generation facilities, vibration sensors continuously analyse frequency spectra, detecting deviations on a micro level that predict mechanical imbalance. If it goes beyond learned tolerance levels, the system can adjust rotational speed or automatically request an inspection without intervention from a human.

From a technological point of view, good performance at the edge for analytics is enabled by the advances of embedded artificial intelligence: miniaturization at the level of the neural network model and the appearance of special hardware accelerators. That means compact AI architecture is included in modern edge processors, such as TinyML, quantized convolutional networks, and spiking neural networks. Compact solutions will be capable of handling complex pattern recognition with a minimum of energy and computational resources. More importantly, embedding intelligence within microcontrollers will supply fields with autonomy at the network's edge, automating sensors themselves, thus allowing their operation to adapt to local conditions. Third, adaptive algorithms enable periodic synchronization with cloud servers, where model parameters are updated using aggregated global knowledge, while local inference and decision-making remain autonomous at the network edge.

The value of edge analytics does not stop at mere latency reduction: it increases both resilience and privacy by first decentralizing computation, making the system resilient against potential network disruptions or cyberattacks against the server. If one of the communication links fails, on the contrary, edge devices continue their operation undisturbed and thus guarantee continuous monitoring for safety. Similarly, in parallel with local processing, sensitive operation or biometric data needs only to be sent outside the device as already aggregated metrics, hence protecting privacy. This fully aligns with both cybersecurity standards, IEC 62443, and the principles of privacy by design in human-centric industry settings. Thereby, in organizations operating in energy, mining, or pharmaceuticals, this kind of architecture contributes to ensuring compliance while keeping IP and worker data integrity intact.

A big difference that edge analytics brings in is how operators and supervisors do their job: in legacy systems, human experts interpret alarms and decide whether an intervention is needed. In the architecture enabled by the edge, this shifts to a role of strategic oversight and validation. Responses that call for immediate attention will be executed autonomously by local devices. Human supervision ensures consistency, accountability, and adherence to ethics. This is a hybrid intelligence framework wherein responsibility is distributed yet transparent. For their part, operators will interact with visual interfaces showing only aggregated or prioritized alerts. Cognitive overload is reduced, and attention can shift to truly critical events.

Practical applications of edge-based safety analytics exhibit the role of those technologies in changing how industrial risk is managed. Within automotive manufacturing, for instance, distributed edge nodes in real-time monitor robotic arms, welding stations, and conveyor belts, while algorithms detect deviations in electrical current patterns or motion trajectories and immediately warn against mechanical anomalies long before production downtime would occur. On offshore oil platforms, autonomous nodes analyse environmental variables like humidity, salinity, and pressure to prevent corrosion and gas leaks. Nodes operate even when satellite connectivity is poor. In smart steel plants, edge systems conduct acoustic-emission and high-frequency thermal-image analysis from furnaces to provide immediate feedback about process stability and material integrity. A tighter integration of sensing, computation, and actuation in a locally situated node builds a new epistemology of industrial safety – for instance, Wang et al. (2025b), Sharifzada et al. (2025), and Jonek-Kowalska and Wolniak (2022, 2023), in which knowledge is produced and used directly at the source of risk.

One of the challenges in deploying analytics based on edges is related to the distribution and synchronization of the models: AI models, normally trained in the cloud, have to be periodically deployed or updated across hundreds/thousands of devices operating in heterogeneous environments. Maintaining consistency while allowing local adaptation requires sophisticated orchestration mechanisms. That's why techniques such as federated learning only recently became instrumental in this respect: they allow local devices to train their models on site-specific data and share independently with the central system only the learned parameters for improving collective accuracy without compromising data privacy. By adopting this approach, the whole safety network becomes self-evolving; hence, it is able to adapt to contextual variations in different geographical or process conditions.

Another methodological aspect is explainability and traceability of local decisions: since most edge analytics will be working in an autonomous way, their reasoning needs to be interpretable for industrial safety. Explainable AI modules at the edge give a visual explanation of the outcome of anomaly detection, pointing out which set of parameters fired an alert and how those signals differed from the learned baseline. It casts more light on and reinforces accountability for post-incident investigation and continuous improvement of the system. This would help meet relevant ethical and legal expectations about automated decision-making that involves critical infrastructure environments, discussed among other things by Ding et al. (2025) and Macrae (2025).

The predictive maintenance ecosystems likewise depend on edge analytics. It enables the scrutiny of sensory data without latency and, hence, the estimation of component degradation in real time. Wind farms are a good example: each individual turbine has an edge processor attached to it, continuously analysing vibration and torque to anticipate, weeks in advance, when it needs servicing. Results are then passed to the supervisory system, where predictive scheduling algorithms coordinate service operations with weather forecasts and projections of energy demand. This combination of local prediction and central orchestration gives expression to the mix of autonomy and coordination intrinsic to modern cyber-physical systems.

They are a part, then, of the broader paradigm of sustainable digitalization for industrial ecosystems – they reduce energy consumption in both communication networks and data centres because of the limitation of data transit as much as possible. They extend machinery life by limiting material waste due to the avoidance of accidents and optimization of maintenance. Because of local data processing, they reduce the need for cloud infrastructures, usually characterized by high carbon footprints. For this reason, the application of edge computing is not only a technological issue but also an ecological responsibility and element of operational sustainability – a fact well captured by Boretti (2025), Zhao et al. (2025), Skotnicka-Zasadzień and Wolniak (2026), and Skubis et al. (2024).

3.3 INTEGRATION WITH SCADA AND PREDICTIVE MAINTENANCE

The integration of intelligent sensing architectures into SCADA systems is indeed an important milestone in industrial safety management. Indeed, for a long period, SCADA systems have formed the backbone of process supervision and operational control in such strategic industries as energy, metallurgy, petrochemistry, and water management. However, the classic architecture of a SCADA system – robust as it is – was designed for deterministic automation and never met the dynamic and adaptive needs of a modern AI-based safety ecosystem. It is precisely this integration of sensor networks with edge analytics and predictive maintenance that brings about this paradigm shift from static control logic to self-learning and self-optimizing systems, capable of anticipating failures and autonomously mitigating risks.

Classically, SCADA collects input from remote terminal units and programmable logic controllers, presenting them on operator consoles and sending field responses through control loops. This model is predictable and reliable only to the extent that the trajectories of the industrial processes themselves are predictable. In reality, an industrial environment is intrinsically stochastic: the vibration pattern changes with wear, temperature varies with ambient conditions, and material properties change along the production cycle. As SCADA incorporates intelligent sensing systems, it is no longer simply a supervisory platform for control; importantly, it is about learning from context. Based on continuous feedback between sensors and local analytics to central databases, SCADA evolves into a coordinating central nervous system, distributing intelligence throughout the plant.

The core of such integration is the ability to provide bidirectional communication between the AI-based sensory layer and the SCADA supervisory layer. Traditional SCADA systems operate largely based on a vertically directed flow: information moves upwards from sensors for visualization and downwards from controllers for actuation. In contrast, modern architecture introduces intelligence exchange horizontally. Real-time executing edge-based AI modules analyse sensor data and send only the important insights – anomalies, predicted degradation indices, or health scores – to SCADA. In turn, SCADA puts this information into context by merging it with process parameters, historical trends, and maintenance records. All that results in an integral view of operational safety, joining the mechanical, environmental, and organizational aspects of the production.

A tangible operational benefit of such integration is predictive maintenance or PdM. Unlike preventive maintenance, which is scheduled at regular intervals regardless of the actual condition of the equipment, predictive maintenance is data-driven, aimed at pinpointing the right time for intervention. Performance indicators refer to temperature gradients, vibration spectra, acoustic emissions, and lubricant chemical composition – each continuously sensed by intelligent sensors. Minor deviations way in advance of mechanical fatigue or component wear are discerned by machine learning algorithms. When the anomaly threshold is exceeded, computation of the Remaining Useful Life estimate is sent to the SCADA interface. The possibility now exists for focused interventions in advance of a critical breakdown with less downtime and lower overall cost. This was also recently demonstrated by Liu et al. (2025a), Alqahtani et al. (2025), and Alzubaidi et al. (2025).

Condition-based monitoring realizes the entire predictive cycle within four steps in the closed loop: data acquisition, condition assessment, prognostic modelling, and then maintenance optimization. Data acquisition shall be done through sensor networks with continuous feeds into edge analytics units. Condition assessment classifies current operational health according to local and global models. Prognostic modelling will be an algorithm, such as survival analysis, random forests, or recurrent neural networks, that makes a prediction of the probability of failure. And lastly, maintenance optimization translates those probabilistic forecasts into actionable schedules, integrated with the planning modules of SCADA. In some more advanced configurations, SCADA shall have the capability to trigger work orders autonomously in computerized maintenance management systems, automatically allocating resources and sending notifications to technicians.

SCADA and predictive analytics, therefore, interact to create a kind of cyber-physical symbiosis: while sensors and edge devices provide the perceptual intelligence of the system, it is SCADA that provides the contextual and executive intelligence. They represent the hierarchical structure of decision-making in which cognition flows upwards as well as downwards. Consider an example in which a vibration sensor on a compressor detects an anomaly. The edge module labels the event as medium risk and relays an alert to SCADA. SCADA, in turn, cross-references the event with its databases of historical maintenance records and load data. If the anomaly coincides with increased temperature and past signs of bearing degradation, the system automatically raises the alert level to critical and optimizes operational parameters with a view to reducing stress on the component. This is a coordinated response within seconds, embodying prevention by its very definition: autonomy (Asma et al., 2025; Hong et al., 2025).

Data interoperability is inseparable from the new paradigm of SCADA. While SCADA applications were based on specific industrial protocols, intelligent sensing systems employ several communication standards – MQTT, OPC UA, Modbus TCP, and REST APIs, to name a few – for data exchange. Such seamless integrations require middleware layers that can translate heterogeneous data formats into uniform semantic structures. Standardization in ontology-based integration frameworks, widely supported by the Industrial Internet Reference Architecture (IIRA), supports this process. Due to semantic interoperability, SCADA will be able to understand what those numeric values mean; for example, it will know that 'VIB_03' is the identifier of a vibration sensor located in turbine section B and operating under load condition class 2. This is highly relevant for intelligent decision-making, reducing the cognitive workload for the human operator.

Beyond predictive maintenance, smart sensors now integrate with SCADA and redefine situational awareness in industrial operations. AI analytics today equips SCADA interfaces to provide operators with real-time visualization of health indices of machines, manpower, and environments. Hazardous zones can be spotted with phenomenal speed through no less than 3D process models, thermal overlays, and dynamic risk maps. For instance, metallurgical plants can be fitted with thermal cameras feeding colour-coded heat-stress zones onto the virtual plant maps via SCADA integration, hence correlating sensor signals with spatial risk patterns in real time. The operator transforms from a reactive observer to an informed decision-maker, aided by machine intelligence and contextual visualization.

This transformation is importantly enabled by the increasing presence of digital twins: virtual duplicates of physical assets that connect with sensor networks and SCADA data. Each of these twins would be regularly updated with actual measurements; this, in essence, would let operators simulate any number of different scenarios of equipment performance and degradation in real time. With such integration, the SCADA operator will be able to virtually test any hypothetical intervention before performing it in reality with a minimum risk of human error. Predictive maintenance algorithms learn from performance in digital twins and feed optimized recommendations back into SCADA. In this way, there is a continuous improvement loop where sensing, modelling, and control form an intelligent ecosystem of perpetual learning.

The capability of such architectures is underlined by various industrial implementations that combine SCADA with AI-based predictive maintenance. In this regard, gas turbine monitoring systems at Siemens Gas & Power use AI-enhanced modules that analyse acoustic and vibration signatures to predict blade fatigue. Besides this, overall fuel efficiency is increased, and operational stability is assured through having automatically scheduled maintenance cycles. Similarly, the Tata Steel Jamshedpur plant couples intelligent sensing data with SCADA-driven predictive models responsible for monitoring furnace temperature uniformity and oxygen concentration with a view to preventing material defects. Further, the Predix platform developed by General Electric is able to act like a hybrid ecosystem of SCADA-cloud, whereby intelligent analytics spot the potential turbine or compressor anomalies several weeks in advance of their actual failure, whereas the creation of digital tickets for maintenance is directly integrated into enterprise systems (Lin et al., 2025; Mohammed & Al-Naib, 2025; Mubarak et al., 2025; Valderrabano-Gonzalez & Castañeda, 2025).

The strategic implications of this convergence go far beyond issues of operational efficiency. First, intelligent sensing through SCADA completely changes the structure of risk governance for the organization in that decision-making is distributed, evidence-based, and temporally dynamic. Safety will depend less on human vigilance than on an ensemble of intelligent agents working cooperatively in real time. Predictive maintenance enabled by AI reshapes the cultural approach from fail-and-repair to predict-and-prevent. It is this epistemological reorientation that enhances resilience and trust in automation and furthers the alignment of industrial safety with the higher ambitions of sustainability and Industry 5.0: fostering human-machine cooperation and systemic adaptability.

As promising as this may sound, the integration process does raise several thorny challenges: first, there is compatibility with legacy systems – most of the SCADA installations are very old, without standardized APIs or enough computational resources to interface with AI-based modules. Retrofitting requires edge gateways that double up as translators and computation nodes. Huge streams of data emanating from predictive maintenance raise serious issues related to data governance: ownership, data retention, and compliance with protection laws on data. Third, cybersecurity – the increased connectivity of SCADA means an expanded attack surface. Security architectures should therefore incorporate AI-driven intrusion detection and encrypted channels of communication. Finally, there are human factors to consider: operators have to be trained to interpret probabilistic predictions and learn how to trust algorithmic insights without losing situational control (Bousla et al., 2025; Verhappen, 2025).

Convergence to unlock these constraints comes from the concept of cognitive SCADA systems in research and industrial practice. These next-generation supervisory platforms embed real-time learning, adaptive visualization, and human-in-the-loop feedback mechanisms. The system will not just conduct runtime logic in cognitive SCADA but will even learn from reinforcement and interaction with operators. An interface which self-adapts to user expertise highlights those anomalies that are most relevant for the ongoing task. Such systems will, over the long term, mediate between fully autonomous AI and human strategic overviews to form the ethical and operational core in intelligent industrial ecosystems.

BIBLIOGRAPHY

Alqahtani, M. H., Shaheen, A. M., & Atiea, M. A. (2025). A meteorology-free wind power forecasting using a hybrid physics-aware, SCADA-only ensemble residual learning. *Results in Engineering*, *28*, 107285.

Alzubaidi, F. Y. H., Kurnaz, S., Naseri, R. A. S., & Farhan, H. M. (2025). SCEN-SCADA security: An enhanced osprey optimization-based cyber attack detection model in supervisory control and data acquisition system using serial cascaded ensemble network. *Optimal Control Applications and Methods*, *46*(5), 2178–2199.

Angel Joseph, P., & Sivaraju, S. S. (2025). Enhancing electric vehicle charging safety and efficiency through hybrid charging systems and intelligent management strategies. *Journal of Energy Storage*, *118*, 116073.

Asma, K., Raja, M. A. Z., Chang, C.-Y., Shu, C.-M., & Shoaib, M. (2025). Machine learning knowledge-driven investigation for immunity-infused fractional industrial virus transmission in SCADA systems. *Journal of Industrial Information Integration*, *48*, 100940.

Boretti, A. (2025). Advanced battery thermal management: A review of materials, cooling systems, and intelligent control for safety and performance. *Energy Storage*, *7*(7), e70273.

Bousla, M., Belfkir, M., Haddi, A., Mourabit, Y. E., & Sadki, A. (2025). Comparative evaluation of Weibull parameter estimation methods for wind energy forecasting: A case study of the Tetouan wind farm with SCADA-based availability integration. *Results in Engineering*, *27*, 106835.

Chen, F., Zhao, Z., Hu, X., Yin, X., & Yang, J. (2025). Enhancing the safety of hydroelectric power generation systems: An intelligent identification of axis orbits based on a nonlinear dynamics method. *Energy*, *324*, 135864.

Cruz-Miranda, J., Damas, M., Al-Soliman, E. N., Díaz, J., & Cappelli, M. (2026). Bridging open and proprietary control frameworks for fusion facilities: EPICS–Industrial SCADA integration in the CODAC system of IFMIF-DONES. *Fusion Engineering and Design*, *222*, 115481.

Cui, T., Li, S., & Wang, X. (2025). Safety intelligent system and overall solution of mine intelligent safety. *Mining Safety and Environmental Protection*, *52*(3), 8–21.

Ding, X., Hasanipanah, M., Monjezi, M., Nguyen, T., & Ulrikh, D. V. (2025). Enhancing mine blasting safety: Developing intelligent systems for accurate flyrock prediction through optimized group method of data handling methods. *Natural Resources Research*, *34*(2), 1199–1218.

El-Nabi, S. A., Ramadan, K. F., El-Rabaie, E.-S. M., Emam, A., & El-Shafai, W. (2025). A real-time design and implementation of intelligent drowsiness and fatigue recognition system for enhancing driver safety. *Engineering Applications of Artificial Intelligence*, *162*, 112665.

Ghadiri, M., & Persio, L. D. (2025). Hybrid SDE-neural networks for interpretable wind power prediction using SCADA data. *Electricity*, *6*(3), 48.

Guo, W. (2024). Research on data privacy protection and heterogeneous data fusion of joint learning in intelligent sensor networks. In *2024 IEEE 2nd International Conference on Sensors, Electronics and Computer Engineering (ICSECE 2024)* (pp. 1570–1574).

Hasrat, I. R., Abdullahi, S. M., & Kang, E.-Y. (2026). Quantitative assessment of energy efficiency, comfort, and safety in an intelligent heating system under false data injection attacks. In *Lecture Notes in Computer Science, 15955 LNCS* (pp. 327–339). Springer.

Herglotz, C., Jabłoński, I., Karnapke, R., Shahin, K., & Reichenbach, M. (2025). Designing intelligent sensor networks: A comprehensive survey of sensing, edge AI, and 5G+/6G connectivity. *IEEE Access*, *13*, 172306–172325.

Hichri, A., Hajji, M., Mansouri, M., Elboughdiri, N., & Gasmi, H. (2025). Enhancing reliability and safety of uncertain grid-connected photovoltaic systems based on intelligent transient regime analysis. *Applied Energy*, *397*, 126231.

Hong, H. S., Hue, N. T., Ninh, N. T., Thuan, N. D., & Huong, N. T. L. (2025). Physics-constrained scheme for outlier removal in wind turbine SCADA data for power curve modeling. *Electric Power Systems Research*, *248*, 111953.

Hu, C., & Lin, H. (2023). Energy conserving and fault tolerant routing algorithm based on backup node set in intelligent sensor network of coal mine. *Chinese Journal of Sensors and Actuators*, *36*(3), 469–474.

Jonek-Kowalska, I., & Wolniak, R. (2021). Economic opportunities for creating smart cities in Poland: Does wealth matter? *Cities*, *114*.

Jonek-Kowalska, I., & Wolniak, R. (2022). Sharing economies' initiatives in municipal authorities' perspective: Research evidence from Poland in the context of smart cities' development. *Sustainability*, *14*, Article 4.

Jonek-Kowalska, I., & Wolniak, R. (2023). *Smart Cities in Poland: Towards Sustainability and a Better Quality of Life?* (p. 200). Routledge.

Jonek-Kowalska, I., & Wolniak, R. (2025). *Exclusion in Smart Cities: Assessment and Strategy for Strengthening Resilience* (p. 224). Routledge.

Kachanun, V., Mueangthongkham, K., Pantaraksakul, P., Tantidham, T., & Hu, C.-L. (2026). Intelligent safety-redundant IoT-edge system for indoor gas leakage detection. In *Communications in Computer and Information Science, 2380 CCIS* (pp. 122–136).

Khadidos, A. O., Selvarajan, S., Alsadhan, N. A., & Singh, S. (2025). CyberSentry: Enhancing SCADA security through advanced deep learning and optimization strategies. *International Journal of Critical Infrastructure Protection*, *50*, 100782.

Kong, F., & Li, J. (2025). Research on intelligent sensor network for laboratory data acquisition and processing. In *Proceedings of the 2025 Asia Europe Conference on Cybersecurity, Internet of Things and Soft Computing (CITSC 2025)* (pp. 277–282).

Kull, K., Khan, M. A., Asad, B., Kallaste, A., & Vaimann, T. (2025). Adaptive deep belief networks and LightGBM-based hybrid fault diagnostics for SCADA-managed PV systems: A real-world case study. *Electronics*, *14*(18), 3649.

Li, L., Zhou, X., Zhang, W., Li, X., & Chen, F. (2025a). Thermal sentinel: Low-earth orbit infrared intelligent system for flying civil aircraft safety. *Remote Sensing of Environment*, *328*, 114826.

Li, Y., Zuo, D., Jin, W., & Qiu, S. (2025b). Intelligent detection of safety hazards along high-speed railway lines based on optical remote sensing images. *Guangxue Xuebao Acta Optica Sinica*, *45*(7), 0728004.

Liang, W. (2025). Energy optimization in intelligent sensor networks: Application of particle swarm optimization algorithm in the deployment of electronic information sensing nodes. *Energy Informatics*, *8*(1), 95.

Lin, K., Yang, S., Zhou, A., & Huang, Y. (2025). Impact of terrain on inflow factors and wind turbine vibrational responses: Insight from SCADA data and wind tunnel tests. *Energy*, *334*, 137712.

Liu, B., Gong, T., Lei, T., Tang, K., & Zhou, Q. (2025a). Design and optimization of intelligent high-altitude operation safety system based on sensor fusion. *Sensors*, *25*(15), 4626.

Liu, D., Ao, Y., Ye, C., Li, J., & Xu, X. (2025b). A study on the inverse identification method of line parameters in distribution networks by fusing SCADA and μPMU data. *Engineering Research Express*, *7*(4), 045202.

Liu, Y., Li, X., Wang, Y., & Li, T. (2024). Optimal layout algorithm design in large-scale civil engineering projects based on intelligent sensor networks. In *Proceedings of the 2024 Asia Pacific Conference on Software Engineering, Social Network Analysis and Intelligent Computing (SSAIC 2024)* (pp. 840–846).

Macrae, C. (2025). Managing risk and resilience in autonomous and intelligent systems: Exploring safety in the development, deployment, and use of artificial intelligence in healthcare. *Risk Analysis*, *45*(4), 910–927.

Mohammed, M. I., & Al-Naib, A. M. T. I. (2025). Design and implementation of a PLC-based SCADA system for real-time monitoring and control of energy conversion. *Journal of Engineering Sciences*, *53*(6), 196–210.

Mubarak, S., Habaebi, M. H., Islam, M. R., Jaleel, N., & Siddique, M. T. (2025). Randomized CNN-based deep learning technique for cyber-attack detection in SCADA industrial control systems. *Measurement: Journal of the International Measurement Confederation*, *254*, 117933.

Nguyen, H. N., & Koo, J. (2025). Enhancing SCADA security using generative adversarial network. *Journal of Cybersecurity and Privacy*, *5*(3), 73.

Philip, A. S., Skaria, A. S., Menon, A. S., Hariprasad, C. K., & Jose, R. (2026). Intelligent fall detection vest with integrated airbag system for elderly safety. In *Lecture Notes in Electrical Engineering, 1368 LNEE* (pp. 57–66).

Pukanec, M., & Tanuška, P. (2026). Design and implementation of a SCADA and ordering system for semi-forms manufacturing. In *Lecture Notes in Networks and Systems, 1558 LNNS* (pp. 381–390).

Rosak-Szyrocka, J., & Wolniak, R. (2025a). *AI-Driven Sustainability: The Future of Human Resources Management* (p. 224). Taylor & Francis Group.

Rosak-Szyrocka, J., & Wolniak, R. (2025b). *Smart Universities in Smart Cities: Shaping the Future of Education and Urban Innovation* (p. 232). Routledge.

Sharifzada, H., Wang, Y., Sadat, S. I., Javed, S., & Khan, S. (2025). An image-based intelligent system for addressing risk in construction site safety monitoring within civil engineering projects. *Buildings*, *15*(8), 1362.

Shiuab, S. S., Venkatasubramanian, V., Jandhyala, V. K., & Torresan, G. (2025). Deduction of oscillation amplitudes using SCADA. *Sustainable Energy, Grids and Networks*, *43*, 101911.

Skotnicka-Zasadzień, B., & Wolniak, R. (2026). Artificial intelligence in public administration. In *Routledge Studies in Innovation, Organizations and Technology* (p. 170). Routledge.

Skubis, I., Wolniak, R., & Grebski, W. W. (2024). AI and human-centric approach in smart cities management: Case studies from Silesian and Lesser Poland Voivodships. *Sustainability*, *16*, 18.

Tabrizchi, H., & Kuchaki Rafsanjani, M. (2025). Cloud, fog, and serverless computing environment for intelligent sensor networks in smart cities. In *Digital Twin and Blockchain for Sensor Networks in Smart Cities* (pp. 17–35). Elsevier.

Tao, Y., Song, W., Zhu, Z., Zhang, S., & Lu, C. (2023). Optimization analysis of power coal-blending model and its control system based on intelligent sensor network and genetic algorithm. *Journal of Sensors*, *2023*, 3898063.

Tomar, B., Kumar, N., & Sreejeth, M. (2025). A smart PLC and SCADA-based control framework for automated assembly line in process industry with effluent ratio control strategy. *Sadhana: Academy Proceedings in Engineering Sciences*, *50*(3), 222.

Valderrabano-Gonzalez, A., & Castañeda, C. E. (2025). Enhanced AI-driven harmonic optimization in 36-pulses converters for SCADA integration. *Electronics*, *14*(18), 3623.

Verhappen, I. (2025). SCADA makes connections. *Control*, *38*(7), 11.

Wali, S., Farrukh, Y. A., Khan, I., & Hamilton, J. A. (2025). Covert penetrations: Analyzing and defending SCADA systems from stealth and hijacking attacks. *Computers and Security*, *156*, 104449.

Wang, J., Topilin, I., Feofilova, A., Shao, M., & Wang, Y. (2025a). Cooperative intelligent transport systems: The impact of C-V2X communication technologies on road safety and traffic efficiency. *Sensors*, *25*(7), 2132.

Wang, S., Qiu, H., He, Y., & Ren, J. (2025b). Research on safety and security co-analysis methods based on the AEB system of intelligent connected vehicles. In *Proceedings of the 2025 4th International Conference on Cyber Security, Artificial Intelligence and the Digital Economy (CSAIDE 2025)* (pp. 58–69).

Wang, W., & Huang, L. (2025). Research on intelligent sensor network technology for mine environment monitoring. *Proc SPIE*, *13447*, 134472A.

Wang, W., Zhu, Q., & Gao, H. (2023). Drift detection of intelligent sensor networks deployment based on graph stream. *IEEE Transactions on Network Science and Engineering*, *10*(2), 1096–1106.

Wei, F., & Yang, Q. (2025). Intelligent sensor network deployment and data analysis in nuclear chemical engineering. *Proc SPIE*, *13574*, 135742V.

Wolniak, R., Gajdzik, B., Grebski, M., Danel, R., & Grebski, W. (2024a). Business models used in smart cities – Theoretical approach with examples of smart cities. *Smart Cities*, *7*, 4.

Wolniak, R., Kowalska-Styczeń, A., Jonek-Kowalska, I., & Michalak, A. (2025). Methods for measuring the effectiveness of human–AI collaboration in an organization. In L. M. Camarinha-Matos, A. Ortiz, X. Boucher, & A. L. Soares (Eds.), *Hybrid Human–AI Collaborative Networks. 26th IFIP WG 5.5 SOCOLNET Working Conference on Virtual Enterprises, PRO-VE 2025, Porto, Portugal, October 27–29, 2025, Proceedings, Part II*.

Wolniak, R., & Stecuła, K. (2024). Artificial intelligence in smart cities: Applications, barriers, and future directions—A review. *Smart Cities*, *7*, 3.

Wolniak, R., Stecuła, K., & Aydın, B. (2024b). Digital transformation of grocery in-store shopping: Scanners, artificial intelligence, augmented reality and beyond—A review. *Foods*, *13*, Article 18.

Xi, J., Zhu, J., Zhang, X., & Guan, L. (2025). Research on work safety intelligent inspection system based on identifier resolution and dynamic encryption algorithm. In *ACM International Conference Proceeding Series* (pp. 424–428).

Ye, T., Chen, H., Liu, G., Li, X., & Zhang, X. (2025a). Advancing the safety of intelligent rail transit systems: A segmentation network for efficient end-of-track degradation feature extraction. *Engineering Applications of Artificial Intelligence*, *159*, 111527.

Ye, T., Chen, H., Liu, G., Li, X., & Zhang, X. (2025b). Erratum to "Advancing the safety of intelligent rail transit systems: A segmentation network for efficient end-of-track degradation feature extraction." *Engineering Applications of Artificial Intelligence*, *158*, 111988.

Zhang, Y., Ran, H., Peng, Y., & Zheng, Y. (2020). A novel framework of artificial intelligent geologic hazards detection over comprehensive remote sensing. *Lecture Notes in Electrical Engineering*, *657*, 457–470.

Zhao, Y., Sun, T., Zhang, H., Lang, X., & Chen, S. (2025). AI-enhanced electrochemical sensing systems: A paradigm shift for intelligent food safety monitoring. *Biosensors*, *15*(9), 565.

Zhou, X. (2024). The real-time monitoring and control of intelligent sensor networks in electrical automation. In *Proceedings of the 2024 5th International Conference on Computer Engineering and Intelligent Control (ICCEIC 2024)* (pp. 188–191).

Part II

Modern Applied Technologies and Industrial Solutions

4 Wearable Technology and Worker-Focused Safety Monitoring

4.1 FALL DETECTION, FATIGUE MONITORING, AND BIOMETRIC FEEDBACK

Current occupational safety management systems face a profound transformation influenced by the Safety II and Safety III paradigms, which no longer focus on reaction to incidents but shift the attention towards anticipation of dangers, comprehension of daily work, and reinforcement of organizational adaptive capacity (Hollnagel, 2018; Leveson, 2016). From this vantage point, wearables – fall detection devices, fatigue monitoring gadgets, and biometric feedback machines – are not merely sensors but incorporated pieces of the organizational risk perception puzzle. They extend the organization's capacity to signal the weak signals, interpreting worker load, dynamics of operations and creating operational knowledge in near real time – one of the foundations for today's quality and safety management systems (Al-Hashimy & Yao, 2025; Meng et al., 2024; Naranjo et al., 2025).

A key driver in this evolution is smart digital systems, which have been described by the European Agency for Safety and Health at Work (EU-OSHA) as 'connected technologies able to monitor work environments, human behaviours and exposure of workers, and to analyse this data to provide operational and strategic decision-making' (EU-OSHA, 2025; Meng et al., 2024; Purushothaman et al., 2025; Zhang et al., 2025b). These include wearables, smart environmental sensors, localization systems, and analytical artificial intelligence (AI) among other tools to offer real-time feedback (Purushothaman et al., 2025). Such an ecosystem will not only be a measuring infrastructure but also represent an organizational capability transcoding biometric, non-biometric, and contextual records into actionable information about real work processes as they have emerged above overloads states that are invisible to traditional supervision models (Nioata et al., 2025).

Empirical evidence clearly indicates to the increasing organizational pertinence of wearables (Kakhi et al., 2025). As the OSH Pulse 2022 survey shows, only 11% of workers in EU use wearables during work, 19% work where digital sensors control conditions, and 7% have health parameters controlled (like heart or posture) (EU-OSHA, 2025). Data also reveal that take-up is markedly higher in large organizations with advanced quality-management systems, underlining the growing trend towards integrating wearable technologies within organizational strategies rather than as separate technical projects (EU-OSHA, 2025).

DOI: 10.1201/9781003754213-6

One of the most established fields of applications is fall detection. Accidents caused by slips, trips, and falls are one of the most serious in terms of workplace risks; in 2019, there were almost 600,000 such accidents reported throughout the EU with more than 500 deaths (EU-OSHA, 2025). Systemic theories highlight that it is not just 'behaviour' that causes falls, but the interplay of environmental, ergonomic, and organizational elements. For example, wearables (smart insoles, inertial measurement unit [IMU] vests, helmets, including impact sensors or footwear feature gait pattern analysis) sense micro-variations of balance, overloading events, and postural instability. The literature demonstrates that the model based on IMU using machine leaning classifiers achieves a classification accuracy of more than 90% for fall and dangerous posture recognition (Naranjo et al., 2025). EU-OSHA (2025) also records fall detection devices with an integrated communication facility, for example, smart insoles which allow a worker to send discrete alarm messages when he falls to the ground. Data from these systems inform 'hot spot' analysis and the redesign of workplace layouts based on design-for-safety considerations (EU-OSHA, 2023). Table 4.1 presents examples of wearable fall detection technologies.

A second important area in the quality and safety management is the monitoring of fatigue (Fiegler-Rudol et al., 2025; Kakhi et al., 2025; Umer et al., 2025). Today, the fatigue in literature is not dependent on an individual lack but reflects a systemic

TABLE 4.1
Examples of Wearable Fall Detection Technologies

Technology	Description	Organizational Relevance and Safety Impact
Smart insoles with pressure sensors and micro-accelerometers	Smart insoles provide real-time monitoring of plantar pressure changes, fine foot micro-movement, and major postural change. They can sense slips, trips, loss of balance, and falls and also let workers send a silent distress alert by tapping their feet together.	Enables fast emergency responses, near-miss identification, and the visualization of spatial risk maps which guide how the workplace should be redesigned according to 'design for safety'.
Inertial measurement units (IMUs) embedded in workwear	IMUs track acceleration, angular velocity, and body orientation. The machine learning models are able to identify hazardous movements and detect falling with very high precision.	Facilitates the application of predictive safety measures by revealing patterns of instability. Facilitates Safety II-based proactive interventions.
Fall detection system within helmets or safety vest	Sensors measure excess force of impact and irregular change in shape or posture; the x4 automatically transmits a warning message with real-time location.	Decreases response time at large sites (such as refineries and mines). Reduces underreporting as it records accident involvements automatically.
Smart Boots – analysing the gait and preventing slipping	Boots can analyse gait, sense whether a surface is slippery, and identify subtle biomechanical deviations that are signs of trouble.	Helpful for dangerous terrain; can detect dangerous conditions and worker overexertion. Aids organizations in prioritizing maintenance and redesign of workflow.

(Continued)

TABLE 4.1 (CONTINUED)

Technology	Description	Organizational Relevance and Safety Impact
NB-IoT + GPS enhanced fall detection and location tracking	Features high-precision fall detection (FSM+IMU) paired with continuous location tracking via NB-IoT/GPS that reports to the cloud for a confidence of up to 99.7% classification accuracy.	Critical for dispersed industrial sites. Produces a fast localization and enables data-driven incident management.

Source: Own study based on Alghayadh et al. (2025), Aznar-Gimeno et al. (2021), Liu et al. (2020), Mishra et al. (2022), Peters and Wischniewski (2023), Ravikumar et al. (2024), and Tseng et al. (2025).

consequence of an overload of work (Ma et al., 2025; Pan et al., 2025). The Work to Zero initiative explicitly claims that 'fatigue has been identified as an underlying, or systemic, risk factor in serious injuries and fatalities' (NSC, 2020; Zwetsloot et al., 2020). The programme mentions that over 43% of US workers experience chronic sleep deprivation, particularly those in high-demand occupations, and has estimated that the organizational costs impacted by fatigue range from USD 1,200 to 3,100 per worker per year (NSC, 2020). Wearables allow heart-rate variability (HRV), electrodermal activity (EDA), balance oscillations, micro-movements, muscle engagement, and quality of sleep – factors closely related to cognitive and physical fatigue – to be monitored. EU-OSHA (2023) reports their use in the transport, mining, and construction sectors to detect microsleeps, decreased vigilance, and induced delayed reaction times. Implementation evidence is promising, with a reduction of 86% in fatigue events within 60 days when fatigue-alert notifications were implemented from the notes reported in the Driver Safety System (NSC, 2020), demonstrating the ability of wearables to translate physiological strain into quantifiable organizational indicators for use within work-system design. Table 4.2 presents examples of wearable technologies for fatigue monitoring.

A third major area of application is biometric feedback – real-time indicators provided directly to workers that help the worker in decision-making, assist them to avoid unsafe postures, or encourage ergonomic self-reflection (Killoran et al., 2025; Persiani et al., 2021). Among these are haptic, visual, and audible cues, directional evacuation signals, smart helmets with augmented-reality displays, and AI-powered feedback algorithms. Haptic cues – vibrations produced by wearables equipment in response to hazardous movements – are more efficient especially when working in noisy or visually restricted environment (Lind et al., 2020; Patel et al., 2025; van Wegen et al., 2023). Research confirms that such signals induced immediate corrections in posture and decreased unsafe postures by 30–50% (EU-OSHA, 2023; Naranjo et al., 2025). With the aid of biometric feedback, worker discretion is increased, reliance on external surveillance diminished, and a safety culture in keeping with Just Culture principles (learning, trust, and shared responsibility) is further underpinned (Dekker, 2017, 2018; Sebastian et al., 2025). Table 4.3 presents examples of wearable technologies for biometric feedback.

TABLE 4.2
Examples of Wearable Technologies for Fatigue Monitoring

Technology	Description	Organizational Relevance and Safety Impact
Biometric wristbands and chest straps (HRV, EDA, and heart rate)	Record HRV, electrodermal activity, stress markers, and temperature allowing an early detection of physical and cognitive fatigue.	Allow to schedule rest and rotations pre-emptively and turn fatigue into an organizational metric instead of personal characteristic.
IMUs for gait-based fatigue detection	Capture subtle changes in balance, diaper length, and acceleration which indicate the onset of advancing fatigue.	Detect fatigue before workers notice symptoms; aid the evaluation and redesign of ergonomic processes.
Advanced sleep-monitoring wearables	Monitor sleep length, cycles, and microsleeps; gather data for fatigue risk assessment between shifts.	Critical for shift work, transport, logistics, and mining; reduce the number of fatigue-related incidents.
Smart biometric vests	Check for heat stress and metabolic fatigue by observing respiration rate, body temperature, muscle tension, and hydration.	Delivers real-time physiological load monitoring for a more effective work-rest ratio and enhanced safety in high-heat conditions.
Eye-tracking augmented-reality wearables	Monitor eye gaze patterns, blink rate, dwell time, and microsleeps – all critical signs of cognitive fatigue.	Enhance situational awareness and minimize attentional failures in high-risk tasks (e.g., machinery and energy).
Industry 4.0 ergonomic wearables (work-related musculoskeletal disorder prevention)	Sensors monitor body posture, joint load, repetitive movement, and muscle fatigue to avoid musculoskeletal disorders.	Convert ergonomic risk into a quantifiable value for long-term process optimization.

Source: Own study based on Aznar-Gimeno et al. (2021), Cerqueira et al. (2020), Gao et al. (2024), Hwang et al. (2025), Machado et al. (2025), Manta et al. (2020), Mathunjwa et al. (2025), and Yoon and Choi (2023).

TABLE 4.3
Examples of Wearable Technologies for Biometric Feedback

Technology	Description	Organizational Relevance and Safety Impact
Haptic smart garments	Clothes emit vibrational signals when they sense poor posture or overexertion, giving instant warning to straighten up and ease off.	Enhance self-regulation, and diminish overdependence on supervisory control; cultivate a proactive safety mentality.
Light- and sound-based feedback systems	Wearables flash warning lights or beep when a worker exceeds fatigue thresholds, or takes on potentially dangerous postures.	Elevate risk awareness and assist in-the-moment behavioural adaptation.

(Continued)

TABLE 4.3 (CONTINUED)

Technology	Description	Organizational Relevance and Safety Impact
Haptic evacuation guidance systems	Vibrational beacons guide workers to exits in emergency situations without visual or audible cues.	Enhance safety of evacuation and mitigate cognitive loading in emergencies.
AR smart helmets with environmental sensors	Give heads-up display alerts for gas, temperature, noise and structural dangers.	Improve the effectiveness of situational awareness and facilitate informed, rapid decision-making.
AI-powered exoskeletons	Exoskeletons monitor the workers' movements and supply counteracting haptic feedback to avoid overloads and dangerous postures.	Prevent musculoskeletal injury and promote ergonomic learning for life.
AI real-time feedback systems	Embedded algorithms process biometric and environmental information to produce actionable micro-recommendations (e.g., 'slow down' and 'take a break').	Develop trust and shared responsibility using Safety II-defined adaptive feedback loops.
Sempulse Halo biosensor (2024)	Miniaturized wearable adhesive biosensor for continuous heart rate, SpO_2, respiration, temperature, motion, and shock monitoring.	Delivers high-fidelity biometric understanding even in harsh environments (mining, emergency response, and military).

Source: Own study based on Guo et al. (2023), Merchán-Cruz et al. (2025), Patel et al. (2025), Rashed et al. (2025), Satheesh et al. (2025), Shen (2025), Slater et al. (2025), Sun et al. (2025), van Wegen et al. (2023), and Jaju and Sulam (2024).

4.2 WEARABLES-TO-SAFETY DASHBOARD INTEGRATION

Connected-worker technologies, ranging from wearables and real-time location systems (RTLS) to analytics platforms that have been integrated into broader Internet of Things (IoT) safety infrastructure solutions, are now highly relevant in the modern digital safety ecosystem (Patel et al., 2022; Siegmann et al., 2025).

The global connected-worker market was valued at USD 4.84 billion in 2023 and is expected to reach USD 14.82 billion by 2031, a clear and burgeoning expansion of solutions developed to support workers operating in high-risk industrial environments, according to the Connected Worker Market Report (2023).

It has been proven empirically that wearable devices have tangible safety benefits. Organizations that have deployed the MakuSafe system – which combines wearable sensors with an analytics platform – report a 60% drop in workers' comp costs and a 50% reduction in recordable incidents within 2 years of rollout (MākuSafe, n.d.). PwC's assessment also stresses MakuSafe's role in reducing insurance premiums and increasing efficiency by enabling the early spotting of hazards and taking advantage of automatic reporting workflows (PwC, 2023).

There is also recorded proof that overlaying wearable sensing with AI-driven analytics can influence worker behaviour. The examples presented in Sustainable Business Magazine include wearable sensors and AI technology that contributed an

84% decrease in unsafe body postures and a 50–60% reduction in the number of injuries to help validate the importance of these systems to ergonomic risk prevention. Scientific assessments within industrial ergonomics also show that modern motion-tracking wearables can provide accuracies as high as 95%, whereas machine learning models can obtain up to 98.6% in fatigue and posture deviation detection (Sustainable Business Magazine, 2023).

The magnitude of the challenge solved by connected-worker solutions is enormous. In 2022, manufacturing, warehousing, and construction accounted for over 700,000 non-fatal injuries and over 2,000 occupational fatalities; meanwhile musculoskeletal injuries alone incurred at least USD 17.7 billion in employer-paid costs (U.S. Government Accountability Office, 2024a and 2024b, 2025). In this light, some of the evidence referenced by the American Society of Safety Professionals in finding that workplace injuries have declined by approximately 20% with wearables discussed is indicative of their actual impact on safety culture (American Society of Safety Professionals, 2022).

The wearable devices have been the basic application of today's industrial safety management system. Originally being developed for physiological tracking and positioning, wearables have become essential data platforms feeding into increasingly complex analytics systems – such as safety dashboards – to support in real-time decision-making processes, risk assessment, and prediction (Maia et al., 2022; Nioata et al., 2025).

Safety dashboards are integrated digital interfaces that collect, process, and represent information from different industrial systems such as Supervisory Control and Data Acquisition (SCADA), Building Management Systems (BMS), or Enterprise Resource Planning (ERP). The integration of these data streams allows companies to move from a reactive to predictive safety approach. Such integration at multiple levels provides the ability to comprehend comprehensively and make sense of the physical, physiological, and operational conditions that affect workforce safety (Budde et al., 2022; Rath & Mishra, 2025). Figure 4.1 presents layered architecture of an integrated system for industrial safety management through wearable technology. The model formalizes the top-down data integration flow from sensor-level (at-the-edge) data collection to higher level analytics and decision support in a safety dashboard.

Today's wearables, such as IMU, GPS/UWB positioning, EDA, and HRV sensors, are uploaded to centralized platforms in real-time fashion. These data streams are transformed into event using algorithms based on events which can in turn either trigger exceptions for instance through an alert mechanism, when the analysis concludes to an anomaly or enrich automated decision logic embedded on dashboards (Mojumder, 2025). In high-risk areas, dashboards may warn against warning signs of fatigue and suggest a rest break, reallocation of tasks, or emergency intervention if they are crossed (Nioata et al., 2025).

An interesting property of integrated dashboards is that they have the capability to correlate multi-source data in real time. For instance, considering an employee's physiological stress signs in combination with environmental aspects (e.g., heat and noise) or occupational information (e.g., overtime from ERP), dashboards may project personalized risk levels and suggest preventive measures. This methodology is beyond traditional compliance monitoring and towards adaptive safety governance (Maia et al., 2022).

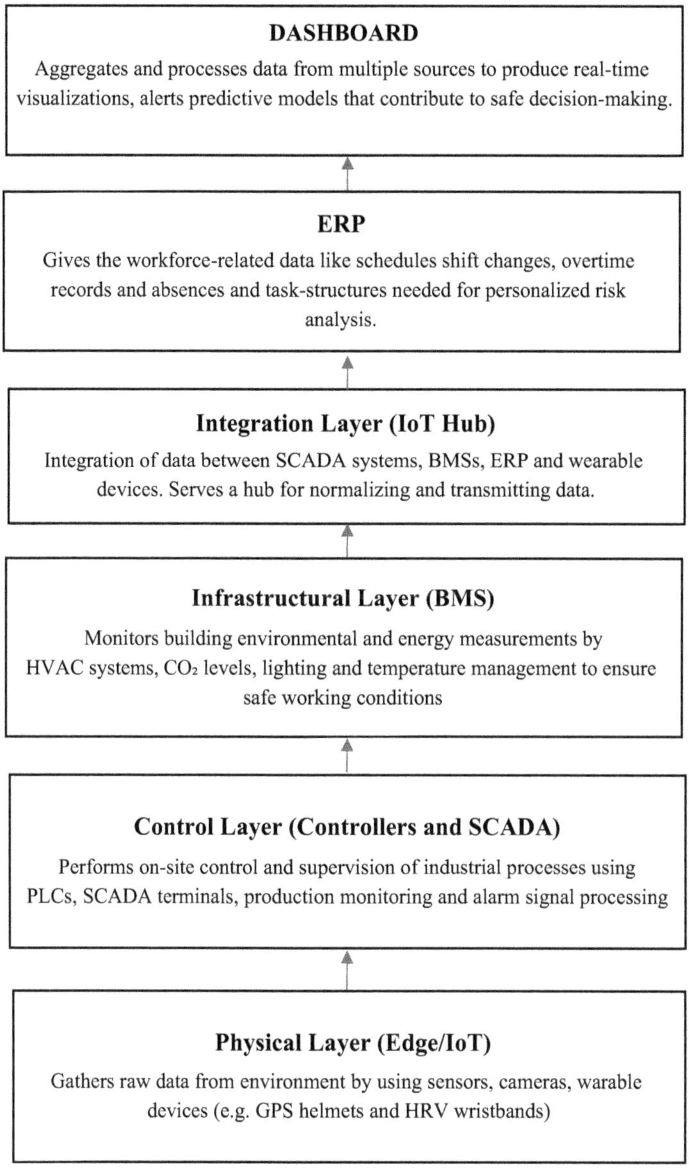

FIGURE 4.1 Wearables-to-safety dashboard integration architecture.

(Own study based on Budde et al. (2022) and Maia et al. (2022).

These systems are increasingly being complemented by AI techniques. Prediction engines, for example, support vector machines, decision trees, and recurrent neural network 'learn' from past occurrence to predict events such as falls, overexertion, or procedural anomaly (Rath & Mishra, 2025). These algorithms improve the predictive resolution of the dashboard and enable enhanced control on a per-dynamic work environment basis.

Digital twins push the capabilities of the dashboard even further by simulating workers and their environments, both physically and behaviourally. This simulation concept provides a method to predict dangers in simulated situations and in the design of safer processes. According to Bahfir et al. (2025), real-world applications of digital twins in manufacturing have shown the betterment in pre-incident detectability as well as system-level safety analytics.

The value of dashboards for compliance and auditing should not be underestimated. Safety dashboards that are integrated into the solution are able to export structured data sets according to standards such as ISO 45001, complete with detailed traceability of safety events for internal and regulatory investigations. According to Budde et al. (2022), dashboards document deviations, incident response times, and exposure histories which contribute to a wider organizational memory.

In addition to sensing, dashboards are using more and more ERP solutions for profiling behavioural risk and job-task exposure. Linking biometric stress markers with ERP-induced data related to shift duties, overtime and team assignment enable to develop adaptive work schedules in order to minimize occupational stress in long term and increasing resilience (Bahfir et al., 2025; Mojumder, 2025). Table 4.4 presents wearables-connected safety dashboard functions.

There is a growing literature around wearable technology and 'connected-worker' systems as part of safety architecture for high-risk environments, including the chemical and petrochemical industry (Patel et al., 2022). The Spot-r case study from implementation at INVISTA chemicals factories offers a very illustrative report on demonstrating first empirical evidence that there are real practical advantages of linking RTLS technologies with processes and safety actions (Chong et al., 2023b).

To prime for a large turning operation at the INVISTA Victoria, Texas (USA) site, hundreds of Spot-r wearables, about 650 equipped worker wearable devices, and critically operational asset EquipTags around 100 were installed to monitor operations (United Rentals, n.d.). It was fit to permanently monitor the field locations of workers over an area of approximately 4,500 acres with a high level of data granularity and real-time access.

There were few tangible gains from the implementation. For one, automating the processes when employees enter and exit the building would put an end to old-fashioned, mistake-prone sign-in/sign-out methods. The location data also have facilitated identifying operational behaviour-specific patterns such as unnecessary movements across zones, choke points at check points, or the time taken in tool rooms. These data were used for management to implement organizational changes which led to better operational performance. As a result, the plant had achieved 63% efficiency on its turnaround project – more than the industry average of around 50% (United Rentals, n.d.). This finding underlines the powerful effect of digital process visibility on job performance in a highly complicated industrial setting.

Also significant was the effect on worker safety. The automated fall detection feature and accurate indoor staff location accelerated the response time in urgent cases. Managers responded to evacuations and were quickly able to check who or who was not accounted for at muster points, significantly reducing the requirement for hand searching and exposure to potential secondary impacts. These findings further validate previous literature documenting that wearables aid emergency response

TABLE 4.4
Wearables-Connected Safety Dashboard Functions

Dashboard Function	Description
Real-time physiological monitoring	The dashboard augments data from HRV, EDA, and IMU sensors included in wearable devices for psychophysical monitoring of workers; because, it supports early indication against stress, overload, and waning concentration.
Environmental hazard visualization	The dashboard analyses data from HRV, EDA, and IMU sensors embedded in wearable devices so that the psychophysical status of workers can be monitored. It is an early diagnosis for stress, overload, or poor concentration.
Fatigue prediction	Machine learning techniques scrutinize biometric changes over time to forecast the development of critical fatigue levels. Facilitates early action with Rest Advisories or task reassignment.
Behavioural risk profiling	The dashboard correlates ERP data (i.e., overtime shift rota and absent reasons) with behavioural markers generated by the wearables to determine individual risk exposure level and/or maximize workers' assignment.
Geofencing and zonal alerts	Based on GPS/UWB data, the system can detect when a worker enters restricted or dangerous zones or stays too long in areas of risk. The contextual cues (e.g., Vibration or Audio feedback) are provided in real time.
Incident logging and auditability	Record events, anomalies, alarms, and user responses in standard formats (e.g., csv and xml) as required by occupational safety regulations such as ISO 45001 to enable internal and external audits.
SCADA/BMS/ERP data fusion	Wearable data can be contextualized (e.g., correlated) with SCADA readings (e.g., vibrations of machines), BMS readings (HVAC failures), and ERP readings (workload), which alert the system to systemic hazards or overload.
Digital-twin-based simulations	To simulate such situations (e.g., high-temperature exposure during ventilation failure), a virtual model is utilized in which the worker and work environment are modelled, making it possible to plan ahead of time before a true incident happens.
Automated safety alerts	The worker and work environment are modelled in the virtual space to simulate, for instance, a high-temperature exposure due to ventilation breakdown making it feasible to make preparations for real incidents.
Compliance reporting and retrospective analysis	Dashboard generates periodic reports from historical data, which can help in preventive safety measures, HSE performance reviews, and the submission of compliance documents to regulatory authorities.

Source: Own study based on Aksüt and Eren (2025), Assaad et al. (2025), Maloney et al. (2025), Mudiyanselage et al. (2025), Pang et al. (2020), Umer et al. (2025), Wei et al. (2023), and Zhang et al. (2025a).

and the detection of procedural breaches and risk-related behavioural signals (Chong et al., 2023a; Nnaji et al., 2021).

The INVISTA case shows that wearable technologies can provide important contributions to safety management systems as well as add value for the operational efficiency, work organization, and resource management. Results suggest that by combining spatial and behavioural data with the management process, operational practices can contribute to ongoing productivity increases and greater adherence to procedures – and hence this could be an important outcome in highly variable environments where there is a high presence of operational risk.

4.3 WORKER PRIVACY, ETHICS, AND REAL-TIME RESPONSE

The incorporation of wearable devices, location-based systems, and analytical approaches seems to be one of the main trends for digitalized OSH management. According to the European Agency for Safety and Health at Work (EU-OSHA), 'smart digital systems' offer the opportunity to identify hazards earlier, gain a better understanding of working conditions, and take a more proactive approach to prevention but also pose new risks in terms of data protection, surveillance, and psychosocial issues for workers whether they are co-workers or other users, such as clients or citizens who are being monitored (EU-OSHA, 2019, 2022). From a safety management standpoint, this implies that connected-worker solutions cannot simply be considered as technical artefacts as they are part of larger socio-technical systems on which trust, fairness, and the legitimacy of monitoring-practice concerns are just as important as hazard-detection features.

Sensor-based systems and wearable devices are an especially sensitive aspects of digital safety (e.g., they can monitor environmental values such as temperature and gas concentration) in conjunction with biometric signals such as heart rate, HRV, electrodermal activity (EDA), movement patterns, fatigue, and physical load (Moon & Ju, 2024). Studies show that industrial wearable IoT allows for an ongoing, situational assessment of risk and can be used to predict injuries stemming from overexertion, too much ambient exposure, or sudden health deterioration. However, as highlighted in Okonkwo et al. (2025), these systems also bring about great risks concerning data security and profiling, with high potential for unauthorized access to sensitive worker health details.

The literature consistently emphasizes how biometric data are particularly privacy sensitive as such data can provide information on health, status issues, as well stress response and behavioural patterns (Doherty et al., 2025). Many users of these devices are worried about the secondary use of their data not just by manufacturers but also by employers and insurers. Such challenges range from the inference of an individual's health pre-dispositions and a person's relative performance (e.g., due to fatigue or stress) all the way up to their social interactions based on motion and location information. So, it is essential that the amount of data used from wearable sensors must not go beyond the requirement to maintain healthy working environment, thus following an ethical wearable deployment should rely on principles of minimizing useless data (Doherty et al., 2025; EU-OSHA, 2019).

Transparency and consent are issues occupying at the heart of the industrial wearables discussion. Svertoka et al.'s (2021) review suggests that workers' consent improves when companies deploy clear rules on data collection, storage, and dissemination and when employees understand how sensors, algorithms, or data-processing routines work. Non-transparency and stealthy use of wearable technologies undermine trust, provoke resistance, and have privacy intrusion implications (Ball, n.d.). Most workers are willing to agree to surveillance if the data are used only for the protection of health and safety rather than performance evaluation, discipline, or micromanagement (Moon & Ju, 2024).

An important ethical issue is the behavioural consequences of wearables. There is evidence that warning devices, which delivered haptic, visual, or auditory alarms, could alter the pace of work and patterns of movement and rest. Moon and Ju (2024) stress that employees react favourable to these systems when the warnings are not too frequent, uncontrollable, or coercive. Too much signalling can result in alarm fatigue, a condition in which users no longer respond to alerts over time. This mechanism is well-documented in the literature on clinical alarm systems and IoT devices and has also direct implications for industrial setting: an overflow of low-priority alarms leads workers to ignore severe ones.

The study on electronic monitoring in the workplace also demonstrates that the technology may influence psychosocial dimensions, stress, perceived pressure, loss of autonomy, and fear of evaluation (Ball). In industries where the work is physically demanding, if continuous monitoring of physical load becomes wearable and normalized, it also risk getting associated with performance measures rather than as a technique to ensure safe procedures. Hence, the principle of proportionality – a favourite in legal and ethical discussions – suggests that wearable data are to be used only as far as it is needed for the prevention of hazards (EU-OSHA, 2019).

A second important aspect is the interoperability between wearables and industrial systems. In cyber-physical systems research, researchers emphasize that the integration of biometric data with environmental, production, and location data generates very detailed worker profiles which are fertile ground for predictive analytics but also pose challenges for ownership and control over data (Budde et al., 2022). Okonkwo et al. (2025) reported that in industrial environment, wearable IoT is highly susceptible to cybersecurity issues such as data eavesdropping, unauthorized access, or identity reconstruction from biometric streams. Decentralized architecture, data anonymization, and preventing the transmission of sensitive signals are identified as being key to subverting most effective data privacy solutions reported in literature.

With the increasing number of wearable devices, the problem of safety culture is also becoming more prominent. The Just Culture and Safety II principles, extensively discussed in the field of occupational safety (Dekker, 2017; EU-OSHA, 2022), suggest that data should not be employed to blame individuals but rather to aid organizational learning. These models claim that data should not be employed to blame – and thus discipline – workers but rather to support an organization's learning (Dekker, 2017). They propose moving away from a punitive discourse to an understanding of the human–technology–organization relations that inform critical events. For real-time systems, this implies considering risk detection algorithms that create knowledge about how to improve rather than punish people for not following procedures.

Real-time response is both the best and most challenging aspect of connected-worker technologies. Success of real-time interventions is found to be contingent upon proportionality, contextual relevance, and whether users know why or when the system acts (Moon & Ju, 2024). Any interventions, particularly the ones caused by bad data, trigger-happy thresholds, or opaque algorithms, can damage trust and degrade system performance. Table 4.5 presents enhanced ethical, privacy, and real-time response issues of connected-worker technologies.

TABLE 4.5
Enhanced Ethical, Privacy, and Real-Time Response Issues of Connected-Worker Technologies

Technology	Risk and Ethics Description
Biometric wearable sensors (HRV, EDA, fatigue indicators, and posture monitoring)	Long-term physiological and movement data can provide a detailed view on health status, stress indicators, fatigue levels, physical constraints, as well as long-term behavioural patterns. This data also permit the development of more comprehensive health and behavioural profiles for workers. Dangers of moving from safety-based monitoring to biometric tracking. Displacing data from preventive use to performance evaluation compromises both proportionality and worker autonomy.
RTLS systems (UWB, RFID, and GPS)	Real-time tracking offers activity patterns, time spent in zones, co-worker interactions, micro-break, and work pace. This could result in more ubiquitous surveillance outside of safety-critical situations. The ethical conflict arises when location information gathered for evacuation situation and security purposes is transformed into a tool used for punishment, discipline, or performance control. Proportionality becomes a key concern.
Smart helmets and smart PPE (gas, temperature, fall, and impact sensors)	Integrate environmental data and activity; with the addition of video, accounts for procedural deviations and team dynamics. Detailed work pattern reconstruction is made possible. Conflict between alerts as protective mechanism and possible interpretability of them as a reprimand. Excessive alarms generate informational overload. Ethical issue: data are meant to inform learning, not punishment.
AR technologies and smart glasses (e.g., RealWear 2020)	Devices may include audio–video, field of view, and occasionally gaze direction. These metrics can uncover thinking, attention, or task-execution across third parties. Risk of violating 'cognitive privacy'. Potential for systematic abuse based on recordings. Chronic apprehension of taking errors in recording stymies learning and risk taking when experimenting.
Worker digital twins	When put together with biological, environmental, processing, and positional data, these profiles are dense and identifiability linked in a way that makes them hard to anonymize. Models can expose work load, stress response, long-term capacity, or risk factors. Potential for behavioural and health profiling, algorithmic inscrutability, decisions based on predictive models that workers cannot comprehend or challenge. Ethical consideration: imbalance in information between institution and observed workers.
Haptic and corrective systems (vibration alerts and posture signals)	Continuous kinematic tracking uncovers work dynamics, pace, resting behaviour, and compensatory techniques. Information could indirectly be used to monitor performance or deviation from the normal behaviour. Ongoing kinematic monitoring provides work style, pacing, rest pattern, and adaptive strategy information. Indirectly, a collected data can be used to identify how well the system is functioning or whether its behaviour deviates from expected ones.
Real-time analytics platforms (integration of SCADA, ERP, BMS, and wearables)	Multi-source fusion presents high-detailed profiles of worker behaviour, physiology status, location, and performance in technological workflow. Risk of a 'completely transparent worker' by deep data integration. Ethical challenge: dividing line between safety analysis and productivity scoring. Programmed decisions feedback organizational biases, while they are opaque and non-transparent as for fairness.

Source: Own study based on Adão Martins et al. (2021), Aksüt and Eren (2025), Chong et al. (2023a), Pang et al. (2020), Slater et al. (2025), Slovák et al. (2021), Soltanmohammadlou et al. (2019), and Wei et al. (2023).

REFERENCES

Adão Martins, N. R., Annaheim, S., Spengler, C. M., & Rossi, R. M. (2021). Fatigue monitoring through wearables: A state-of-the-art review. *Frontiers in Physiology*, *12*, 790292. https://doi.org/10.3389/fphys.2021.790292

Aksüt, G., & Eren, T. (2025). Determination of wearable technological devices according to their use in improvement of health and safety in the mining sector. *Safety Science*, *184*, 106746.

Alghayadh, F. Y., Jena, S. R., Gupta, D., Singh, S., Bakhriddinovich, I. B., & Batla, Y. (2025). Dynamic data-driven resource allocation for NB-IoT performance in mobile devices. *International Journal of Data Science and Analytics*, *19*(4), 659–673. https://doi.org/10.1007/s41060-023-00504-7

Al-Hashimy, H. N. H., & Yao, J. (2025). The moderating effect of internet of things and wearable technologies on enhancing safety management in construction sites. *Construction Innovation*. https://doi.org/10.1108/ci-10-2024-0318

American Society of Safety Professionals. (2022). *Connected Worker Technologies and Workplace Injury Reduction*. ASSP. https://www.assp.org

Assaad, R. H., Mohammadi, M., & Poudel, O. (2025). Developing an intelligent IoT-enabled wearable multimodal biosensing device and cloud-based digital dashboard for real-time and comprehensive health, physiological, emotional, and cognitive monitoring using multi-sensor fusion technologies. *Sensors and Actuators A: Physical*, *381*, 116074.

Aznar-Gimeno, R., Labata-Lezaun, G., Adell-Lamora, A., Abadía-Gallego, D., del-Hoyo-Alonso, R., & González-Muñoz, C. (2021). Deep learning for walking behaviour detection in elderly people using smart footwear. *Entropy*, *23*(6), 777. https://www.mdpi.com/1099-4300/23/6/777

Bahfir, S., Belkorchi, W., & Zemmouchi-Ghomari, L. (2025). A survey of digital twin implementations for industrial transformation: Analysis of real-world case studies. *Australian Journal of Multi-Disciplinary Engineering*, 1–39.

Ball, K. (n.d.). *Electronic Monitoring and Surveillance in the Workplace: Literature Review and Policy Recommendations (No. 9789276414803)*. European Commission. https://research-repository.st-andrews.ac.uk/handle/10023/32743

Budde, L., Hänggi, R., Friedli, T., & Rüedy, A. (2022). Smart factory navigator. In *Smart Factory Navigator* (pp. 7–31). Springer International Publishing.

Cerqueira, S. M., Da Silva, A. F., & Santos, C. P. (2020). Smart vest for real-time postural biofeedback and ergonomic risk assessment. *IEEE Access*, *8*, 107583–107592.

Chong, H.-Y., Wang, X., & Wang, J. (2023a). The adoption intentions of wearable technology for construction safety. *Buildings*, *13*(11), 2747. https://doi.org/10.3390/buildings13112747

Chong, H.-Y., Xu, Y., Lun, C., & Chi, M. (2023b). The adoption intentions of wearable technology for construction safety. *Buildings*, *13*(11), 2747. https://www.mdpi.com/2075-5309/13/11/2747

Dekker, S. (2017). *The Field Guide to Understanding 'Human Error'*: CRC Press.

Dekker, S. (2018). *Just Culture*. CRC Press.

Doherty, C., Baldwin, M., Lambe, R., Altini, M., & Caulfield, B. (2025). Privacy in consumer wearable technologies: A living systematic analysis of data policies across leading manufacturers. *npj Digital Medicine*, *8*(1), 363. https://www.nature.com/articles/s41746-025-01757-1

EU-OSHA. (2025). *Smart Digital Systems and Worker Safety*. European Agency for Safety and Health at Work.

European Agency for Safety and Health at Work. (2019). *Digitalisation and occupational safety and health (OSH): An EU-OSHA research programme*. Publications Office of the European Union. https://doi.org/10.2802/559587

European Agency for Safety and Health at Work, Spyridopoulos, K., Kilic, L., Battaglini, M., Olausson, N. et al. (2022). *Smart digital monitoring systems for occupational safety and health – Workplace resources for design, implementation and use – Report*/ Publications Office of the European Union. https://doi.org/10.2802/315859

Fiegler-Rudol, J., Lau, K., Mroczek, A., & Kasperczyk, J. (2025). Exploring human-ai dynamics in enhancing workplace health and safety: A narrative review. *International Journal of Environmental Research and Public Health*, *22*(2), 199. https://www.mdpi.com/1660-4601/22/2/199

Gao, L., Wang, C., & Wu, G. (2024). Wearable biosensor smart glasses based on augmented reality and eye tracking. *Sensors*, *24*(20), 6740. https://www.mdpi.com/1424-8220/24/20/6740

Guo, K., Orban, M., Lu, J., Al-Quraishi, M. S., Yang, H., & Elsamanty, M. (2023). Empowering hand rehabilitation with AI-powered gesture recognition: A study of an sEMG-based system. *Bioengineering*, *10*(5), 557. https://www.mdpi.com/2306-5354/10/5/557

Hollnagel, E. (2018). *Safety–I and Safety–II*: CRC Press.

Hwang, S., Kwon, N., Lee, D., Kim, J., Yang, S., Youn, I., et al. (2025). A multimodal fatigue detection system using sEMG and IMU signals with a hybrid CNN-LSTM-attention model. *Sensors*, *25*(11), 3309. https://www.mdpi.com/1424-8220/25/11/3309

Jaju, V., & Sulam, S. (2024). Smart helmets. In *Handbook of Artificial Intelligence and Wearables* (pp. 286–298). CRC Press.

Kakhi, K., Jagatheesaperumal, S. K., Khosravi, A., Alizadehsani, R., & Acharya, U. R. (2025). Fatigue monitoring using wearables and AI: Trends, challenges, and future opportunities. *Computers in Biology and Medicine*, *195*, 110461.

Killoran, J., Manseau, J., Park, A., & Kietzmann, J. (2025). Supporting and humiliating dignity with biometric technologies: An affordance perspective. *Journal of Business Ethics*, *200*(2), 307–324. https://doi.org/10.1007/s10551-024-05889-4

Leveson, N. G. (2016). *Engineering a Safer World*. The MIT Press.

Lind, C. M., Diaz-Olivares, J. A., Lindecrantz, K., & Eklund, J. (2020). A wearable sensor system for physical ergonomics interventions using haptic feedback. *Sensors*, *20*(21), 6010. https://www.mdpi.com/1424-8220/20/21/6010

Liu, S., Zhang, J., Zhang, Y., & Zhu, R. (2020). A wearable motion capture device able to detect dynamic motion of human limbs. *Nature Communications*, *11*(1), 5615. https://www.nature.com/articles/s41467-020-19424-2

Ma, P., Pan, C., Shen, H., Shen, W., Chen, H., Zhang, X., et al. (2025). Monitoring nap deprivation-induced fatigue using fNIRS and deep learning. *Cognitive Neurodynamics*, *19*(1), 30. https://doi.org/10.1007/s11571-025-10219-z

Machado, A., Ferreira, D. F., Ferreira, S., Almeida-Antunes, N., Carvalho, P., Melo, P., et al. (2025). A systematic review of chest-worn sensors in cardiac assessment: Technologies, advantages, and limitations. *Sensors*, *25*(19), 6049. https://www.mdpi.com/1424-8220/25/19/6049

Maia, E., Wannous, S., Dias, T., Praça, I., & Faria, A. (2022). Holistic security and safety for factories of the future. *Sensors*, *22*(24), 9915. https://www.mdpi.com/1424-8220/22/24/9915

MākuSafe. (n.d.). *MākuSafe Case Studies: Improving Safety Outcomes with Wearable Technology*. https://www.makusafe.com

Maloney, E. M., Corsi, S. R., Pronschinske, M. A., DeCicco, L. A., Nott, M. A., Frisch, J. R., et al. (2025). Derivation and characterization of environmental hazard concentrations for chemical prioritization: A case study in the Great Lakes tributaries. *Environmental Toxicology and Chemistry*, *44*(7), 2029–2047.

Manta, C., Jain, S. S., Coravos, A., Mendelsohn, D., & Izmailova, E. S. (2020). An evaluation of biometric monitoring technologies for vital signs in the era of Covid-19. *Clinical and Translational Science*, *13*(6), 1034–1044.

Mathunjwa, B. M., Kor, R. Y. J., Ngarnkuekool, W., & Hsu, Y.-L. (2025). A comprehensive review of home sleep monitoring technologies: Smartphone apps, smartwatches, and smart mattresses. *Sensors*, *25*(6), 1771. https://www.mdpi.com/1424-8220/25/6/1771

Meng, F., Cui, Z., Guo, H., Zhang, Y., Gu, Z., & Wang, Z. (2024). Global research on wearable technology applications in healthcare: A data-driven bibliometric analysis. *Digital Health*, *10*, 20552076241281210.

Merchán-Cruz, E. A., Moveh, S., Pasha, O., Tocelovskis, R., Grakovski, A., Krainyukov, A., et al. (2025). Smart safety helmets with integrated vision systems for industrial infrastructure inspection: A comprehensive review of VSLAM-enabled technologies. *Sensors*, *25*(15), 4834. https://www.mdpi.com/1424-8220/25/15/4834

Mishra, P., Pai, P., Singh, P., Kayande, V., & Parmar, M. (2022). Implementation of a smart helmet with alcohol and fall detection and navigation system. In *International Conference on Innovative Computing and Communications, 1388* (pp. 239–251). https://doi.org/10.1007/978-981-16-2597-8_20

Mojumder, M. U. (2025). Impact of lean six sigma on manufacturing efficiency using a digital twin-based performance evaluation framework. *ASRC Procedia: Global Perspectives in Science and Scholarship*, *01*(01), 343–375. https://global.asrcconference.com/index.php/asrc/article/view/16

Moon, J., & Ju, B.-K. (2024). Wearable sensors for healthcare of industrial workers: A scoping review. *Electronics*, *13*(19), 3849.

Mudiyanselage, U. H., Gonzalez, E. L., Sermet, Y., & Demir, I. (2025). An immersive hydroinformatics framework with extended reality for enhanced visualization and simulation of hydrologic data. *Applied Sciences*, *15*(10), 5278. https://www.mdpi.com/2076-3417/15/10/5278

Naranjo, J. E., Mora, C. A., Bustamante Villagómez, D. F., Mancheno Falconi, M. G., & Garcia, M. V. (2025). Wearable sensors in industrial ergonomics: Enhancing safety and productivity in Industry 4.0. *Sensors*, *25*(5), 1526. https://doi.org/10.3390/s25051526

National Safety Council. (2020). *Work to Zero: Fatigue Detection Technologies*. National Safety Council.

Nioata, A., Ţăpirdea, A., Chivu, O. R., Feier, A., Enache, I. C., Gheorghe, M., & Borda, C. (2025). Workplace safety in Industry 4.0 and beyond: A case study on risk reduction through smart manufacturing systems in the automotive sector. *Safety*, *11*(2), 50. https://www.mdpi.com/2313-576x/11/2/50

Nnaji, C., Awolusi, I., Park, J., & Albert, A. (2021). Wearable sensing devices: towards the development of a personalized system for construction safety and health risk mitigation. *Sensors*, *21*(3), 682. https://doi.org/10.3390/s21030682

Okonkwo, C., Awolusi, I., Nnaji, C., & Akanfe, O. (2025). Privacy and security of wearable internet of things: A scoping review and conceptual framework development for safety and health management in construction. *Computers & Security*, *150*, 104275.

Pan, Y., Khodaei, Z. S., & Aliabadi, F. (2025). Online fatigue crack detection and growth modelling through higher harmonic analysis: A baseline-free approach. *Mechanical Systems and Signal Processing*, *224*, 112167.

Pang, Y., He, Q., Jiang, G., & Xie, P. (2020). Spatio-temporal fusion neural network for multiclass fault diagnosis of wind turbines based on SCADA data. *Renewable Energy*, *161*, 510–524.

Patel, S., Rao, Z., Yang, M., & Yu, C. (2025). Wearable haptic feedback interfaces for augmenting human touch. *Advanced Functional Materials*, *36*, 2417906.

Patel, V., Chesmore, A., Legner, C. M., & Pandey, S. (2022). Trends in workplace wearable technologies and connected-worker solutions for next-generation occupational safety, health, and productivity. *Advanced Intelligent Systems*, *4*(1), 2100099.

Persiani, S. G. L., Kobas, B., Koth, S. C., & Auer, T. (2021). Biometric data as real-time measure of physiological reactions to environmental stimuli in the built environment. *Energies*, *14*(1), 232. https://www.mdpi.com/1996-1073/14/1/232

Peters, M., & Wischniewski, S. (2023). Workwear with loosely coupled IMU sensors for posture classification during assembly tasks: A pilot study. In *International Conference on Digital Human Modeling* (pp. 90–97). Cham: Springer Nature Switzerland.

Purushothaman, M. B., Rotimi, F. E., Samarasekara, S., & GhaffarianHoseini, A. (2025). SMART technologies that influence construction health and safety factors risk reduction in the current digital era. *Smart and Sustainable Built Environment*. https://doi.org/10.1108/SASBE-09-2024-0400

PwC. (2023). *IoT Wearable Boosts Workplace Safety and Efficiency: MākuSafe Case Study*. PwC US. https://www.pwc.com/us/en/library/case-studies/makusafe-iot-wearable-technology.html

Rashed, A., Abdulazeem, Y., Farrag, T., Bamaqa, A., Almaliki, M., Badawy, M., & Elhosseini, M. (2025). Toward inclusive smart cities: Sound-based vehicle diagnostics, emergency signal recognition, and beyond, *Machines*, *13*(4), 258. https://www.mdpi.com/2075-1702/13/4/258

Rath, K. C., & Mishra, B. K. (2025). Next-generation computing technology for electric vehicle manufacturing – concept, challenges and future research. *Next-Generation Computing Systems and Technologies*, *1*(1), 1–10.

Ravikumar, S., Babu, R., Kirubasri, G., Vijay, K., & Sanjai, S. (2024). Smart shoes: Empowering fall prevention for the elderly. In *Artificial Intelligence and Information Technologies* (pp. 144–150). CRC Press.

RealWear. (2020). *Improving Safety and Efficiency with Assisted-Reality Headsets: Industrial Case Studies*. RealWear Inc. https://www.realwear.com

Satheesh, N., Gopisankar, N., Kumarganesh, S., Anthoniraj, S., Saravanakumar, S., Sagayam, K. M., et al. (2025). Advanced AI-driven emergency response systems for enhanced vehicle and human safety. *Iran Journal of Computer Science*, *8*(2), 441–456. https://doi.org/10.1007/s42044-025-00228-w

Sebastian, R. A., Ehinger, K., & Miller, T. (2025). Do we need watchful eyes on our workers? Ethics of using computer vision for workplace surveillance. *AI and Ethics*, *5*(4), 3557–3577. https://doi.org/10.1007/s43681-025-00726-4

Sempulse Corporation. (2024). *Sempulse Halo Non-Invasive Vital Signs Monitor: Technical Overview and Specifications*. Sempulse Corporation. https://sempulse.com/non-invasive-vital-signs-monitor/

Shen, B. (2025). Research on synchronous control of light and sound in dance performance based on adaptive algorithm. *Journal of Computational Methods in Sciences and Engineering*, *25*(6), NP34. https://doi.org/10.1177/14727978251371169

Siegmann, K. A., Ivošević, P., & Visser, O. (2025). Working like machines: Technological upgrading and labour in the Dutch agri-food chain. *Work, Employment and Society*, *39*(1), 202–225.

Slater, A. E., Ferris, S., Dixon, J. A., Renshaw, C., Moore, T. E., Frady, G., & Harrison, S. J. (2025). Navigating in zero-visibility: A haptic guidance system for improving egress and situation awareness of professional firefighters. *Human Factors*, *67*(11), 1152–1169.

Slovák, J., Melicher, M., Šimovec, M., & Vachálek, J. (2021). Vision and RTLS safety implementation in an experimental human-robot collaboration scenario. *Sensors*, *21*(7), 2419. https://www.mdpi.com/1424-8220/21/7/2419

Soltanmohammadlou, N., Sadeghi, S., Hon, C. K., & Mokhtarpour-Khanghah, F. (2019). Real-time locating systems and safety in construction sites: A literature review. *Safety Science*, *117*, 229–242.

Sun, L., Lee, B. G., & Chung, W.-Y. (2025). Enhancing fire safety education through immersive virtual reality training with serious gaming and haptic feedback. *International Journal of Human–Computer Interaction*, *41*(9), 5607–5622.

Sustainable Business Magazine. (2023). *How AI-Powered Wearables Are Transforming Worker Safety*. https://sustainablebusinessmagazine.net/business-review/ai-sensors-and-wearables-preventing-workplace-accidents/

Svertoka, E., Saafi, S., Rusu-Casandra, A., Burget, R., Marghescu, I., Hosek, J., & Ometov, A. (2021). Wearables for industrial work safety: A survey. *Sensors*, *21*(11), 3844. https://www.mdpi.com/1424-8220/21/11/3844

Tseng, C.-K., Huang, S.-J., & Kau, L.-J. (2025). Wearable fall detection system with real-time localization and notification capabilities. *Sensors*, *25*(12), 3632. https://www.mdpi.com/1424-8220/25/12/3632

U.S. Government Accountability Office. (2024a). *Technology Assessment: Wearable Technologies for Worker Safety* (GAO-24-106014). Government Accountability Office.

U.S. Government Accountability Office. (2024b). *Science & Tech Spotlight: Wearable technologies in the workplace* (GAO-24-107303). https://www.gao.gov/products/gao-24-107303

U.S. Government Accountability Office. (2025). *Wearable technologies: Potential opportunities and deployment challenges in manufacturing and warehousing* (GAO-25-107213). https://www.gao.gov/products/gao-25-107213

Umer, W., Mehmood, I., Qarout, Y., Antwi-Afari, M. F., & Anwer, S. (2025). Deep learning-based fatigue monitoring of construction workers using physiological signals. *Automation in Construction*, *177*, 106356.

United Rentals, Inc. (n.d.). *Location Services Boost Turnaround Safety and Productivity* [Case Study]. https://www.unitedrentals.com/our-company/case-studies/digital-solutions/location-services-boost-turnaround-safety

van Wegen, M., Herder, J. L., Adelsberger, R., Pastore-Wapp, M., van Wegen, E. E. H., Bohlhalter, S., et al. (2023). An overview of wearable haptic technologies and their performance in virtual object exploration. *Sensors*, *23*(3), 1563. https://www.mdpi.com/1424-8220/23/3/1563

Wei, Z., Liu, K., Liu, X., Li, Y., Du, L., & Gao, F. (2023). Multilevel data-driven battery management: From internal sensing to big data utilization. *IEEE Transactions on Transportation Electrification*, *9*(4), 4805–4823.

Yoon, H., & Choi, S. H. (2023). Technologies for sleep monitoring at home: Wearables and nearables. *Biomedical Engineering Letters*, *13*(3), 313–327. https://doi.org/10.1007/s13534-023-00305-8

Zhang, J., Li, R., Dong, L., Ke, Y., Liu, C., Pei, M., et al. (2025a). Ultrasensitive biodegradable piezoelectric sensors with localized stress concentration strategy for real-time physiological monitoring. *Chemical Engineering Journal*, *507*, 160521.

Zhang, X., Li, J., Zhou, J., Zhang, S., Wang, J., Yuan, Y., et al. (2025b). Vehicle-to-everything communication in intelligent connected vehicles: A survey and taxonomy. *Automotive Innovation*, *8*(1), 13–45. https://doi.org/10.1007/s42154-024-00310-2

Zwetsloot, G., Leka, S., Kines, P., & Jain, A. (2020). Vision zero: Developing proactive leading indicators for safety, health and wellbeing at work. *Safety Science*, *130*, 104890.

5 AR-Enhanced Operational and Safety Support

5.1 AUGMENTED INTERFACES AND FIELD-OF-VISION ALERTS

Augmented interfaces are interaction surfaces that project digital information onto the operator's real space, generally using augmented reality (AR) glasses or safety helmets with AR support (Azuma, 1997; Billinghurst et al., 2015; Craig, 2013). They make it possible to directly display process variables, commands and background information at the point on which the user is focusing. In the scientific literature, they are referred to as perceptual augmentation and safety augmentation because they increase workers' situational awareness and enable the generation of field-of-view alerts – dynamic warnings presented in real time directly within the field-of-view of workers (Cheng et al., 2022; Yu et al., 2024). Field-of-view alerts are fundamentally different from classical static safety signals (signs, stickers and protection fences); they are customized in real time and take into account the context, the actions being carried out by a worker, and the immediate environment that may change according to the machine's operation (Hussain et al., 2022; Rao et al., 2022). The literature, therefore, conceptualizes warnings as a component of safety augmentation – a technology-driven enhancement of human perception for safety through intelligent and adaptive warnings (Winkler et al., 2025; Jamson et al., 2008; Schwarz & Fastenmeier, 2017; Zheng et al., 2004).

AR has been one of the instrumental technologies that has matured towards ensuring safety and operational performance in high-risk sectors like construction, energy, heavy industry and industrial maintenance. The related concept of augmented interfaces – systems that overlay digital layers directly within the operator's field-of-view – has emerged as central to this flow action. They enable context-sensitive data, actions with procedures, warnings and real-time alerts to be present right where the operator's visual focus is instead of having to avert their eyes. Because of this, they have been progressively thought of as perceptual and safety-extension devices that embed dynamic field-of-view alerts into the digital work area (Fernández-Moyano et al., 2025).

Based on the systematic reviews, industrial AR solutions can be associated with the following capability goals: reduce errors as system-guided assistance; improve situation awareness by offering real-time contextual information; ensure procedural compliance, shorten task execution time and address human factors, for example, health and safety (Flor-Unda et al., 2023; Li et al., 2018; Morales Méndez & Del Cerro Velázquez, 2024a; Tatić & Tešić, 2017). Through the use of in situ instructions and digital sensory data overlaid in the physical space, field-of-view AR interfaces juxtapose information from IoT sensors, technical documentation, CMMS systems

DOI: 10.1201/9781003754213-7

and 3D models with the viewed object. In doing so, AR shortens the perception–decision–action loop; all information is in a common perceptual channel, and it is no longer necessary to switch focus back and forth between real-world surroundings and external displays (Belsito et al., 2024).

Experimental research also supports AR in terms of operator performance (Di Valentina et al., 2021; Maida et al., 2007; Yin et al., 2020). In manufacturing settings, researchers find that it is possible to improve workers' efficiency and to reduce manual operation time errors with in situ instructions in the worker's line of sight rather than paper-based manuals or fixed HMI displays (Aygün et al., 2025; BELSITO et al., 2024). These findings are supported by a meta-analysis conducted by Morales Méndez and Del Cerro Velázquez (2024a; 2024b), which found that not only do AR systems enhance objective task quality, but they also have a positive impact on workers' subjective perceived control.

Recent studies showed that the most effective AR solutions are those linked to knowledge management systems and predictive analytics (Hasan et al., 2023; Rosati et al., 2023; Zong & Guan, 2025). Augmented reality, in conjunction with process data, building information model (BIM) models and diagnostic tools, is a unified information channel for the operator to control cyber-physical systems. From this standpoint, the AR interface is not just a visual overlay, but it becomes a mediating layer that enables the operators' orientation and decision-making in complex industrial environments (Bonci et al., 2019; Xie et al., 2022; You & Feng, 2020).

For safety, field-of-vision alerts are vital (Kang et al., 2025; Thorslund & Strand, 2016). AR allows the localization and personalization of safety signals; rather than physical barriers or signs, workers can receive dynamic, location-specific alerts – virtual danger areas, live component warnings (such as an energized part warning), PPE prompts or alerts activated based on machine parameters being out of tolerance. This is one of the key safety augmentation applications where AR enhances hazard recognition, assists decision-making and mitigates human error (Li et al., 2018; Qeshmy et al., 2019; Zong & Guan, 2025).

In those maintenance applications, this function is enhanced by the fusion with data coming from IoT sensors. When temperatures, vibrations or process values exceed pre-defined safety limits, AR devices show colour-coded 'status indicators', procedural instructions or are locked out altogether to indicate the need for evacuation. It has been shown that such solutions can help in early anomaly detection, assist less experienced technicians and minimize unplanned downtime (Compan et al., 2023; Mao et al., 2025; Mayáns-Martorell, 2013; Moghaddam & Klumpp, 2025; Rabhi et al., 2021).

Results presented in the Mao et al. (2025) study also confirm that AR overlays provide means for early error detection and real-time rectification, something that can lead to a direct enhancement of task success rates. Visual augmentation enables precise, localized instruction right where the technician is working, greatly improving safety and quality of work performed. The conclusions are further supported by industrial deployment results. In the energy industry, wearable helmets with AR technology dedicated to remote expert support and field-of-view service inscriptions have reportedly reduced downtime by 20% in offshore wind turbine maintenance (AIIEM, 2024). Within the construction industry, a pilot implementation of the

DAQRI Smart Helmet at Mortenson Construction showed minimized assembly errors, better schedule compliance and enhanced situational awareness by overlaying BIM directly over the actual construction (Chai et al., 2019; DAQRI, 2016a, 2016b).

In the area of workplace safety training AR can be applied, especially when employees need to become familiar with actual equipment (Li et al., 2018; Söbke et al., 2025; Tatić & Tešić, 2017). Research (Ismael et al., 2024) indicates that AR overlays allow workers to better understand hazardous situations and make them confident when applying procedures, including experienced operators. Studies on VR and AR training (Scorgie et al., 2024; Syed et al., 2022) repeatedly underscore the role of AR to be crucial whenever it is necessary to rehearse behaviour within actual operational environments.

At the micro level, AR serves as an 'external working memory', which minimizes cognitive load by offering necessary information directly at the point of performance (Rymarov et al., 2024). At the meso level, AR deployments are integrated into organizational safety management systems – supporting real-time monitoring of state-of-the-art knowledge in terms of the equipment condition, remote engineering supervision and context-aware advice (Solomashenko et al., 2025). The issues with improved interfaces are all about ergonomics (weight of the device, field-of-view, brightness and rendering latency) and organizational matters like costs, data integration, 3D-model quality or training effort. Penczek (Penczek, 2025) underlines a requirement to standardize AR display parameters, particularly when considering content visibility and its influence on spatial perception. Table 5.1 presents layers of an AR field-of-view interface and their roles and application examples in industry.

TABLE 5.1
Layers of an AR Field-of-View Interface and Their Roles and Application Examples into Industry

AR Interface Layer	Role in Industrial Work Environments	Impact on Safety
Sensor layer (Layer A)	Combines information between IoT sensors, diagnostic systems and condition monitoring devices to overlay onto the guided components. With colour-coded markers or geometry overlaid, parameters like temperature, vibration, pressure or overload levels can be seen.	Enabling real-time anomaly detection, improving predictions of blackouts, reducing escalations and speeding response time. Reduces the occurrence of failures stemming from minor or previously unknown parameter changes.
Instruction layer (Layer B)	Provides step-by-step guidance relative to the tasks currently performed by the operator. Guidance appears directly on the part being worked, so workers do not have to look away to reference external documentation.	Reduces mental effort and protects against misinterpretation of procedures or skips. Avoids incorrect manipulations, such as the use of an inappropriate tool, skipping a verification process, and task execution in an improper sequence. Strengthens compliance with safety-critical procedures.

(Continued)

TABLE 5.1 (CONTINUED)

AR Interface Layer	Role in Industrial Work Environments	Impact on Safety
Safety alert layer (Layer C)	Provides real-time context-sensitive alarms: virtual danger zones, moving-equipment alerts, high-temperature warnings, leak status reports or safe-position signals. Alerts change in real time as the environment shifts around the worker.	Improves situational awareness by allowing you to see what you cannot see. Lower the risk of accidents resulting from lack of attention, tiredness or misjudgement. Performs as a safety augmentation, essentially a technology integration for human hazard perception.
Contextual information layer (Layer D)	Offers diagnostic, organizational, and asset-focused information about equipment: equipment identification, fault history, documentation (i.e. how-to manuals), technical drawings and 3-D models linked to a physical asset.	Increases diagnostic efficiency and facilitates data-driven decision-making. It reduces misidentification errors and leads to the enhancement of efficiency and safety in maintenance work.

Source: Own study.

The four AR interface layers are explained individually for the sake of clarity, but actually work as a cohesive perceptual system. These four layers constitute a perceptual stack that ranges from raw sensory cues to task-specific guidance, adaptive safety alerts and higher-level contextual information. This structure parallels cognitive ergonomics literature, which states that operators first respond to immediate sensory cues, then procedural steps and warning alerts, before interpreting contextual reference sources.

5.2 AR FOR INDUSTRIAL TRAINING AND PROCEDURAL ASSISTANCE

Immersive technologies, in particular AR, play a leading role in the training ecosystem and procedural support framework of manufacturing industries shifting from Industry 4.0 to the human-orientated concept of Industry 5.0 (Cabrera-Duffaut et al., 2024; Liberatore & Wagner, 2021; Mallam et al., 2019). Today in practice, AR is finding its way into the operational information infrastructure at an increasing pace – in pilot implementations and early deployments (Alam et al., 2025; Malta et al., 2023). Its main role is to facilitate the merging of physical and digital layers by overlaying instructions, process parameters and warnings that are relevant in a specific context directly onto the visual field of an operator with the aim of increasing his situational awareness and procedural compliance. This expanded workspace serves to bring the process of learning more in line with that of operational execution within a unified perception–decision–action loop ('AI-Driven Sustainability: The Future of Human Resources Management', 2026).

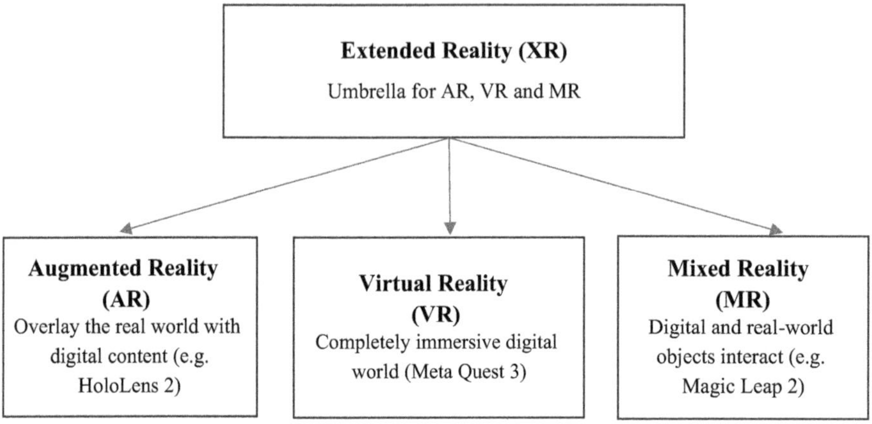

FIGURE 5.1 Components of XR.

(Source: Own study.)

Recent definitions place AR in the larger category of Extended Reality (XR), which includes AR, VR and mixed reality (MR) (Figure 5.1).

Modern XR taxonomies do not only focus on immersion but further include interaction fidelity, context anchoring and system interoperability (Abouhawwash et al., 2024; Hancko et al., 2025; Rauschnabel et al., 2022; Rosak-Szyrocka, 2025; Rosak-Szyrocka et al., 2025; Zywiołek et al., 2024). Opposite to VR, AR keeps continuous access to the real world and provides superimposed, context-relevant information through optical or video see-through head-mounted displays. This perceptual field hybridization allows workers to carry out physical activities with real-time, spatially anchored support – a feature emphasized in recent HMI studies ('AI-Driven Sustainability: The Future of Human Resources Management', 2026; Eswaran, & Bahubalendruni, 2023; Chu et al., 2025; Park et al., 2020; Xia et al., 2024).

In terms of cognitive ergonomics, AR is exploited as an externalized working memory. Experimental investigations in complex industrial work tasks show that superimposing procedural steps and item identifiers into the operator's field-of-view greatly saves cognitive load and avoids switching attention between documentation and real-world objects being manipulated ('AI-Driven Sustainability: The Future of Human Resources Management', 2026; Ariansyah et al., 2022; Mehta, 2016; Wu et al., 2023). These benefits manifest in concrete performance improvements such as quicker task execution, lower rates of error and enhanced procedural chaining. Spatial anchoring of instruction has also been reported to facilitate formation of stable mental models even when the layout of machinery and dependencies is involved (Ma et al., 2025).

An exponentially increasing number of quantitative studies prove that the utilization of AR for industrial training and safety-critical learning creates an efficient learning environment ('AI-Driven Sustainability: The Future of Human Resources Management', 2026; Gualtieri et al., 2023; Hancko et al., 2025; Li & Pilz, 2023;

Stefan et al., 2023). Studies have shown that AR-based safety training generally leads to better performance than traditional modes of training concerning the recognition of hazards, compliance with procedures and retention of information ('AI-Driven Sustainability: The Future of Human Resources Management', 2026; Domgue Kamwa et al., 2024; Gong et al., 2024; Söbke et al., 2025; Zong & Guan, 2025). An automation in construction meta-analysis also concluded that AR training is superior to traditional formats for high-risk environments, as it offers trainees perceptually rich, contextually anchored tasks without exposing them to the risk of real tasks (Hancko et al., 2025; Rana & Chicone, 2025; Yazdi, 2024).

In procedural support, AR allows for the merging of sensor information, context-bound restrictions and step-by-step guidance in a single perceptual channel. A recent work on maintenance engineering shows that AR enhances anomaly detection, assists beginners in diagnostic tasks and cuts unplanned downtime when interfaced with the IoT infrastructures (Costa et al., 2025; Kostadimas et al., 2025; Mao et al., 2025). More evidence also shows the added value of AR when adopted in conjunction with cyber-physical production systems for decision-making support, downtime reduction and inspection task consistency improvement (Bouyahrouzi et al., 2025; Khamaisi et al., 2025). Table 5.2 presents the benefits of AR in industrial training and procedural assistance.

TABLE 5.2
Benefits of AR in Industrial Training and Procedural Assistance

Benefit	Description	Role of AR
Improved training effectiveness and long-term knowledge retention	AR-trained workers have higher retention of procedural knowledge and perform multistep tasks with less variability. Within a very controlled assembly task, AR-educated operators demonstrated even fewer errors, along with a reduction in sequence deviations and better recall following instruction when compared to a more conventional style of instruction.	AR helps minimize distraction by displaying procedural elements directly in the operator's line of sight. This enhances working memory and enables information to be represented in a spatially grounded way, as instructions are mapped onto physical parts interactively.
Enhanced operational efficiency and quality of task execution	A recent meta-analytic review concludes that AR technology decreases task time and both assembly and diagnostic errors across industries. AR allows the users to refer to up-to-date process information, process limits and spatially aligned assistance without breaking their activity.	In AR, procedural advice, parameter bounds and warning prompts are integrated in a perceptual continuum. AR minimizes sequence errors and the search time for information while further preventing holes between the documentation and execution environment, specifications of AR, which are assumed to be achieved by avoiding breaks between documentation and tasks.

(Continued)

TABLE 5.2 (CONTINUED)

Benefit	Description	Role of AR
Support for maintenance operations and real-time diagnostics	AR provides efficiency in anomaly detection and helps diagnostics by contextualizing related information. Augmented reality (AR) for real-time viewing of IoT sensor data, the history of equipment and 3D component models overlaid onto the asset to identify faults more accurately.	AR superimposes virtual information on physical machinery with the goal of integrating cyber-physical system data within the immediate workplace of the operator. This spatial interdependence enables faster diagnosis, better condition monitoring and the ability to make maintenance decisions effectively based on data.
Reduction of cognitive load and stabilization of operator attention	AR efficiently reduces cognitive load by reducing memory load and task interruption. Operators can sustain their attention longer and make fewer errors associated with overload or recollection of a procedure sequence.	AR can serve as an expanded memory tool, liberating cognitive resources to focus on the aspects of decision-making that are task-relevant. By relocating the informational requests from internal memory to a spatially anchored overlay, AR has the potential to lower cognitive load and provide for more sustained attentional engagement.

Source: Own study.

Empirical evidence obtained from industrial deployments makes clear that large operational and cognitive benefits can be achieved when introducing AR-based training environments and procedural assistance systems (Eswaran, & Bahubalendruni, 2023; Baashar et al., 2022; Barros et al., 2024; Daling & Schlittmeier, 2024). Case studies presented in the Boeing, DHL and Capgemini materials demonstrate that AR allows workers to execute tasks while they remain spatially aware and receive contextually relevant information directly within their field-of-view. In this way, AR can aid in bridging the gap between abstract procedural knowledge and on-site task demonstration through placing guidance information within the workspace proper (Barros et al., 2024; BELSITO et al., 2024; Joanna et al., 2024; Rosak-Szyrocka et al., 2023; Schwarz & Fastenmeier, 2017).

One of the widely cited industrial AR applications is Boeing's application of augmented instructions in assembling electrical wiring harnesses. With procedural instructions projected onto the part, workers are no longer required to shift focus between documentation and the physical workpiece.

Boeing states that AR-infused instructional overlays increased task accuracy and advanced recognition of potential mistakes behind the line during pre-production. The company also reports a significant drop-off in time to train, with employees reaching operational proficiency much more quickly than through conventional teaching processes. In the present application, AR served as an on-the-spot training aid to promote correct task performance, contribute to learning speed and increase overall process robustness (The Boeing Company, 2022).

Another illustrative case was in the logistics industry and relates to a pilot deployment of DHL's 'Pick-by-Vision'. Order pickers in the warehouse wore AR smart glasses showing them the way to product locations, item confirmations and guidance on paths to walk.

DHL announced that it saw a greater than 25% increase in warehouse productivity after deploying the solution in a pilot project. Furthermore, AR decreased cognitive demand, reduced identification errors and increased user comfort, with respondents also highlighting the ease with which the system could be learnt. In the light of situated learning theory, these findings indicate that AR helps to provide training where instructional cues are closely linked with task context in logistics operations, thus facilitating the transfer of competence (DHL Customer Solutions & Innovation, 2023; DHL Trend Research, 2014).

The example based in the Capgemini report that surveyed 600 industrial companies on how they use AR and VR. The results also suggest that, for tasks of manual precision (e.g., quality inspection and maintenance support), AR is more mature than VR. In the presented cases, other applications for AR in PCB inspection also enabled improved quality (20–25%) compared with 2D-based techniques, as graphical overlays pointed out parts that will need verification and reduced the risk of misses. And in a third, GE engineers reported performing 59% more service work per hour due to the fact that they could accomplish all of their steps without having to return over and over to a computer or invest time consulting reference materials. The document mentions the application of AR/VR for municipal infrastructure services in TRMUA and how visualizing underground installations increased emergency response efficiency (e.g. during floods, fires or other critical events), as well as technicians working with a virtual model of the terrain without resorting to costly exploratory excavation (Capgemini Research Institute, 2018).

5.3 AR-POWERED REMOTE SUPPORT AND ERROR REDUCTION

Regarding industrial safety, AR is used as a cognitive support fostering intuitive guidance for operators to provide them with context-based, on-the-fly support and to improve the intelligibility of instructions and to reduce errors due to uncertainty when making decisions. Most importantly, allowing arrows, lines, 3D models and animated instructions to be overlaid onto the image of the equipment or infrastructure constituent upon which the technician is working. Spatial markers like these generate a common perceptual space between the remote expert and operator, thereby lessening communication ambiguity and cognitive load – two aspects that have been repeatedly cited in previous research as contributing to human error in safety-critical environments. This way, AR is not just an instructional tool but a medium as well to interpret operational situations (Azuma, 1997). Table 5.3 presents primary categories of industrial AR devices for remote support and error prevention.

Human-centred AI insists that technological systems would be changed far more to match human perceptual limitations and simplify cognitive work when the likelihood of big mistakes remains high. The seminal work of Henderson and Feiner (2011) on AR showed that visual annotations fixed in the visual field of operation improve technical task compliance and minimize errors derived from

TABLE 5.3
Primary Categories of Industrial AR Devices for Remote Support and Error Prevention

Type of AR	Description	Examples
AR smart helmets (head-mounted displays)	A convenient head-worn device for showing instructions and spatial cues into workers natural line of sight while performing the task. Sustain attention with support tasks over procedures using continuous presence. Hands-free use is useful for persons using visual cues or those who have coordination issues with their fingers.	DAQRI Smart Helmet, Trimble XR10
AR smart glasses	Lightweight eyewear that provides visual annotations, remote expert guidance and contextual information at your point of interest. Especially useful in remote expert identification, due to visual communication minimizing the possibility of misunderstanding. Text-based prompting and voice activation provide for ease of use by employees who are deaf or have limited motor control.	RealWear Navigator, Vuzix
Mobile AR devices (Tablets and Smartphones)	Handheld wireless devices, for example, tablets or smartphones for showing augmented instructions, component highlighting or inspection-relevant information. Efficient for inspection and documentation purposes. Accessibility features – zooming, colour and contrast adjustments, screen readers – create usability for staff with vision or cognitive impairments.	iPad (ARKit), ARCore-enabled smartphones
Projection-based AR	Dynamic instruction systems project dynamic instructions onto physical surfaces, being free from having to wear or handhold any devices. Highly accessible, with projected cues easily seen by employees with mobility or sensory disabilities. Reduces the frequency of assembly and sequencing errors by providing steps just where action is.	LightGuide Systems
Adaptive AR	Accessible solutions including intuitive interfaces, typographical optimization, resizing of markers and multimodal feedback (visual, textual and auditory). Provides equivalent access to knowledge about how work activities are performed for employees with sensory, motor or cognitive disabilities. Lowers cognitive burden and improves readability for better decision-making and more accurate task completion.	HoloLens 2, ARKit/ARCore

Source: Own study.

misinterpretation or misexecution of task steps. For example, a worker who is required to locate and extract a particular part in an intricate assembly must often engage in the mental translation of abstract descriptions from word instructions or diagrams when using conventional instructions. AR overcomes this obstacle by indicating the specific element and position via spatial cues. Tang et al. (2002) also reported that this context-sensitive visual feedback reduced execution errors and enhanced precision, resulting in a substantial enhancement of safety during operation (Henderson & Feiner, 2011; Tang et al., 2002).

AR provides a stabilizing force in organizational operations as it diminishes variation attributable to worker experience disparities. Palmarini et al. (2018) argue that context-based information display enables less experienced personnel to work in a way closer aligned with expert behaviour and, hence, promotes process predictability as well as reducing variation from standard practice. Egger et al. (2020) build on this argument, displaying how AR serves as a base that links the single operator to organizational procedural knowledge to counteract the effects of fatigue, inattention and stress. In scenarios where operators have to make quick decisions (e.g. installation, servicing, inspection or emergency actions), minimizing the uncertainty and having a clear view on how to execute step-by-step tasks will further reduce the chance of human errors. Through the step-by-step instructions in AR, attention switching between individual task steps and a picture on work is instant, preventing repetitive activity, which literature has shown to be performance-degrading, as well as unsafe. (Egger & Masood, 2020; Palmarini et al., 2018; Palmarini et al., 2023). Table 5.4 presents the main AR functionalities in remote assistance and the reduction of human error.

TABLE 5.4
Main AR Functionalities in Remote Assistance and Reduction of Human Error

Area of AR Impact	Description	Impact on Safety and Error Reduction
Perceptual support through visual overlays	AR allows accurate generation and display of spatial overlays – for example, arrows, contours, lines, 3D models or animated instructions – on the physical equipment that the operator is interacting with. This serves as a replacement of abstract, 2D instructions with context-anchored guidance in the physical world.	Minimizes the ambiguity and thus reduces cognitive load of translating text or diagrams into spatial acts. Operators perform their tasks with more clarity, precision and in accordance with procedures.
Creation of a shared perceptual workspace	AR enables the remote expert and local operator to share the same cantered visual field in which annotations are locked into physical objects. That common visual context allows for communication in the same 'perceptual language'.	Minimizes misunderstanding of design and contains fewer perspective errors while considering vague verbal directions. Facilitates the right order of workflow and enhances mutual cognition in high-risk procedures.
Reduction of cognitive load	AR relocates some of the operator's cognitive processing from working memory to the visual channel. Instructions are always in context, without the user having to constantly change focus between workspace and documentation.	Decreases vulnerability to mistakes triggered by pressure, stress and fatigue. Helps to keep the performance on a stable level in demanding cognitive workloads.

(Continued)

TABLE 5.4 (CONTINUED)

Area of AR Impact	Description	Impact on Safety and Error Reduction
Equalizing skill levels across workers	Through the use of contextual prompts, step-by-step guidance and spatial queues, AR is enabling novices to work to a fraction of expert levels.	Minimizes variation in task performance and prevents errors as a result of procedure knowledge deficits. Results in more predictable and standardized process results.
Ensuring correct procedural sequencing (workflow enforcement)	AR instructs the worker through a sequence of tasks with clear instructions and visual plans inside the working area. This ensures that no essential steps are missed.	Substantially lowers procedure misses and sequence-related mistakes, especially in high-stakes/urgent domains such as healthcare or energy systems.
Enhancing the precision of remote communication	Annotations and corrections can be added to the live view of the operator by remote experts. It removes vagaries that come with verbal-only and 2D video tutoring.	Minimizes misinterpretation errors and promotes psychological safety with clear, unambiguous confirmation of actions.
Early detection of deviations and design–execution mismatches	AR can enable workers to review the physically constructed environment against digital models (the building information model [BIM]) and detect any breaks in accuracy, as well as omissions or misplacements, before errors become systemic.	Reduces scrap, installation interferences and unsafe conditions caused by parts incorrectly installed or read from sketchy drawings.
Support in high-risk operational environments	Minimizes waste, installation issues and unsafe situations from wrong parts being installed or read from bad prints.	Enhances sense of direction, spatial orientation and decision-making in stressful situations. Helps minimize errors of procedure and judgement in time-sensitive situations.
Strengthening organizational safety culture	AR enhances the transparency of work processes, increases communication and shares informal knowledge less in a company and ensures that the workforce is aligned with common safety standards.	Improves adherence to safety procedures, decreases variability of behaviour, and helps promote a preventive, learning-orientated safety culture.

Source: Own study.

On the aspect of organizational safety, AR also increases the reliability with which users and remote experts communicate. Buń, Grajewski and Górski (Bun et al., 2021) show that the quality of transmission and accuracy of annotations are crucial to keep the operator in a safe state-of-mind condition and to prevent perceptual mismatch-related risks. Conventional voice or video feedback relies on the operator to interpret instructions and apply them to the local context, which leads to potential misunderstanding, ambiguity and errors in sequence. AR helps to cover this interpretive gap, displaying clear, context-embedded cues right into the operator's field of action.

Real industrial deployment evidence shows that AR serves as a cognitive safety margin in the form of part of the mental workload, which emerges when resources in complex tasks are tight. The works of Garcia et al. (2024), found technicians in the railway industry using AR showed increased accuracy of component identification and assembly sequence, subsequently leading to decreased error rates and time taken to carry out tasks.

Recent systematic and scoping reviews of AR in minimally invasive and robot-assisted surgery demonstrate that AR enhances instrument navigation to relevant pathology, increases spatial understanding, enables advanced surgical accuracy and may prevent iatrogenic injury mistakes as well as decision errors in complex processes (Brockmeyer et al., 2023; Chen et al., 2024; Doornbos et al., 2024). Analogous observations are found in procedural training settings, where AR-based step-by-step guidance during ECMO cannulation leads to fewer knowledge-related errors and more adherence to critical steps as compared with conventional training approaches, positioning AR as a perceptual-cognitive augmentation tool informed by the principles of human-centred AI for enhancing safety by increasing perceptual clarity and decision-making capacity (Wang et al., 2017; Wolf et al., 2021).

At the organizational level, integration of AR in work processes has significant impact on cultivating and reinforcing safety culture. Studies on user acceptance in industrial AR have demonstrated that in order for AR to effectively assist operators, its mode of interaction must match the natural work pace and perceptual habits as well as the available digital competencies. However, systematic reviews identify that operator acceptance of AR as safety-improving and control-supporting is determined by reasons such as usability factors, interface transparency and certain cognitive-ergonomic aspects like the reduction of unnecessary visual complexity (Quandt & Freitag, 2021). Workers will be more likely to trust the technology and integrate AR-mediated cues within their habitual safety-related behaviours when AR systems provide information in an intuitive, nonobtrusive, and perceptually integrated manner, reinforcing adherence to formal safety procedures and risk-mitigation norms (Lampropoulos et al., 2024; Quandt & Freitag, 2021).

More evidence also suggests that AR confers its maximum safety benefits through integration into existing organizational safety management structures, which include targeted training programmes, risk-assessment practices and competence-development frameworks. In the context of Industry 4.0 and AR-supported training shows that the effectiveness of AR in terms of operator performance improvements and error reduction grows extensively when adopted as part of procedural systems or socio-technical processes instead of being isolated as a lever (Morales Méndez & Del Cerro Velázquez, 2024a). In such organizational settings, AR serves as a stabilizing influence on the wider socio-technical safety system: it acts to promote consistency in safety-critical practices, to reduce variation in performance due to operators' levels of experience, and to move the organization itself towards a more mature and proactive safety culture that emphasizes ongoing learning and risk understanding (Morales Méndez & Del Cerro Velázquez, 2024b).

CASE STUDY MORTENSON CONSTRUCTION + DAQRI SMART HELMET

The project, by Mortenson Construction and partners DAQRI and Autodesk, marked one of the first applications of AR head-mounted display technology on an active construction site during the build-out for the new Hennepin County Medical Center (HCMC) Speciality Center in Minneapolis. In a pilot implementation, Mortenson employees were able to don the DAQRI Smart Helmet and see three-dimensional BIM models superimposed over the actual construction site, which included in-place mechanical–electrical–plumbing systems, structural components and even forthcoming layers of interior infrastructure. It allowed workers to see a real-time, on-site reality capture of the 'future' state of the building to facilitate cross-discipline coordination and quality control (DAQRI, 2016).

The Smart Helmet incorporated BIM models hosted in Autodesk's BIM 360 cloud service, so workers could 'see-through' wall assemblies and compare the actual construction against design plans. Projecting information such as ductwork or electrical systems onto the physical structure increased comprehension of how it fit together and allowed designers, contractors and medical professionals that help configure clinical space to communicate more effectively (DAQRI, 2016; Levy, 2016).

In terms of functionality, the DAQRI Smart Helmet was designed as an industrial human–machine interface which supports 4D viewing, that is, the spatial overlay of BIM models with temporal and process-related information right inside the field-of-view of the worker. DAQRI marketed the product as a way for field workers to 'see what will be built, see where it needs to be installed, and see how those layers fit together', which is particularly useful in high installation complexity settings such as hospitals or industrial plants (DAQRI, 2015, 2016).

Neither Mortenson nor DAQRI released quantitative statistics on reductions in errors or safety incidents, but embedded within their blog posts was the message that the pilot had led to 'safer and more efficient' construction practices. Workers were able to detect mismatches between the physical environment and the BIM model at an earlier stage and respond faster to installation-related queries (DAQRI, 2016). This usage practice corresponds with established findings in AR construction research that reported that on-site visualization of BIM models helps avoid cross-disciplinary conflict and reduces rework as well as enhances constructability for non-specialist/technologically savvy workers when interpreting complex technical drawings (Nassereddine et al., 2022; Song et al., 2021).

From a safety perspective, the Mortenson use case example demonstrates how AR can contribute to achieving organizational safety goals on reducing human error by increasing informational transparency and early detection of potential execution mistakes. Systematic reviews of AR usage in construction highlight that AR technologies aid design–construction coordination, improve the quality-assured activities, and have the potential to either eliminate unsafe working conditions due to wrongly locating installations or misinterpreting drawings (Nassereddine et al., 2022; Tatić & Tešić, 2017). The Mortenson–DAQRI pilot is one of the few real-world examples to date of AR as a cognitive and perceptual tool in construction on a challenging medical facility project.

REFERENCES

Abouhawwash, M., Rosak-Szyrocka, J., & Gupta, S. K. (2024). *Aspects of Quality Management in Value Creating in the Industry 5.0 Way*: CRC Press.

Aidar Solutions. (2023). *Augmented Reality Remote Assistance: Industrial Applications and Challenges*. Aidar Solutions Technical Report.

AIIEM. (2024). *How Augmented Reality (AR) is Enhancing Industrial Training and Maintenance*. https://aiiem.org/how-augmented-reality-ar-is-enhancing-industrial-training-and-maintenance

Alam, N., Saha, N., Gadow, V., Harik, R., & Ryu, J. (2025). Role of extended reality (XR) technologies in maintenance operations: Trends, challenges, and integration in Industry 4.0. *Manufacturing Letters*, *44*, 1545–1557.

Ariansyah, D., Erkoyuncu, J. A., Eimontaite, I., Johnson, T., Oostveen, A.-M., Fletcher, S., & Sharples, S. (2022). A head mounted augmented reality design practice for maintenance assembly: Toward meeting perceptual and cognitive needs of AR users. *Applied Ergonomics*, *98*, 103597.

Aygün, D., Calepso, A. S., Yang, X., Menges, A., & Sedlmair, M. (2025). Mixing and matching: Instruction conveyance for collaborative tasks using asymmetric augmented reality setups. In *2025 IEEE Conference on Virtual Reality and 3D User Interfaces Abstracts and Workshops (VRW)* (pp. 1072–1078). IEEE.

Azuma, R. T. (1997). A survey of augmented reality. *Presence: Teleoperators and Virtual Environments*, *6*(4), 355–385.

Baashar, Y., Alkawsi, G., Ahmad, W. N. W., Alhussian, H., Alwadain, A., Capretz, L. F., et al. (2022). Effectiveness of using augmented reality for training in the medical professions: Meta-analysis. *JMIR Serious Games*, *10*(3), e32715.

Barros, P. L., Gonçalves, G., & Pinto, V. H. (2024). Augmented reality in industrial training: A comparative analysis of teaching methods. In *International conference on optimization, learning algorithms and applications* (pp. 49–64). Cham: Springer Nature Switzerland.

Belsito, C., Franciosi, C., & Di Pasquale, V. (2024). The impact of augmented reality devices on operator performance in manufacturing contexts. *Management and Production Engineering Review*. https://journals.pan.pl/dlibra/show-content?id=131000

Billinghurst, M., Clark, A., & Lee, G. (2015). A survey of augmented reality. *Foundations and Trends® in Human–Computer Interaction*, *8*(2–3), 73–272.

Bonci, A., Carbonari, A., Cucchiarelli, A., Messi, L., Pirani, M., & Vaccarini, M. (2019). A cyber-physical system approach for building efficiency monitoring. *Automation in Construction*, *102*, 68–85.

Bouyahrouzi, E. M., Benmimoun, R., El Kihel, Y., & Bajjou, M. S. (2025). Integrating industry 4.0 technologies and maintenance 4.0 for sustainable manufacturing: A systematic literature review. *The International Journal of Advanced Manufacturing Technology*, *140*(1–2), 35–61. https://doi.org/10.1007/s00170-025-16194-3

Brockmeyer, P., Wiechens, B., & Schliephake, H. (2023). The role of augmented reality in the advancement of minimally invasive surgery procedures: A scoping review. *Bioengineering*, *10*(4), 501. https://www.mdpi.com/2306-5354/10/4/501

Bun, P., Grajewski, D., & Gorski, F. (2021). Using augmented reality devices for remote support in manufacturing: A case study and analysis. *Advances in Production Engineering & Management*, *16*(4), 418–430.

Cabrera-Duffaut, A., Pinto-Llorente, A. M., & Iglesias-Rodríguez, A. (2024). Immersive learning platforms: Analyzing virtual reality contribution to competence development in higher education – A systematic literature review. *Frontiers in Education*, *9*, 1391560. https://doi.org/10.3389/feduc.2024.1391560

Capgemini Research Institute. (2018). *Augmented and Virtual Reality in Operations: A Guide for Investment*. Capgemini.

Capgemini Research Institute. (2020). *Augmented and Virtual Reality in Operations: Enhancing Efficiency at Scale*. Capgemini SE.

Chai, C., Mustafa, K., Kuppusamy, S., Yusof, A., Lim, C. S., & Wai, S. H. (2019). BIM integration in augmented reality model. *International Journal of Technology*, *10*(7), 1266.

Chen, Z., Cruciani, L., Fan, K., Fontana, M., Lievore, E., Cobelli, O. de, et al. (2024). Towards safer robot-assisted surgery: A markerless augmented reality framework. *Neural Networks: The Official Journal of the International Neural Network Society*, *178*, 106469.

Cheng, J., Wong, P. K.-Y., Luo, H., Wang, M., & Leung, P. H. (2022). Vision-based monitoring of site safety compliance based on worker re-identification and personal protective equipment classification. *Automation in Construction*, *139*, 104312.

Chu, C.-H., Pan, J.-K., & Chen, Y.-W. (2025). Ergonomic workplace design based on real-time integration between virtual and augmented realities. *Robotics and Computer-Integrated Manufacturing*, *92*, 102859.

City University of London. (2022). *ARRA: Augmented Reality for Remote Assistance*. ARRA Project Research Report.

Compan, N., Matias, J., Lutz, M., Rix-Lièvre, G., Brissaud, D., Belletier, C., & Coutarel, F. (2023). The impact of enabling collaborative situations in AR-assisted consignment tasks. *Frontiers in Industrial Engineering*, *1*, 1258241. https://doi.org/10.3389/fieng.2023.1258241

Construction Safety Research Alliance. (2023). *Immersive Technologies for Hazard Identification: Annual Safety Performance Report*. CSRA.

Costa, A., Miranda, J., Dias, D., Dinis, N., Romero, L., & Faria, P. M. (2025). Smart maintenance solutions: AR- and VR-enhanced digital twin powered by FIWARE. *Sensors*, *25*(3), 845. https://pmc.ncbi.nlm.nih.gov/articles/pmc11820622/

Craig, A. B. (2013). *Understanding Augmented Reality: Concepts and Applications*. Newnes.

Daling, L. M., & Schlittmeier, S. J. (2024). Effects of augmented reality-, virtual reality-, and mixed reality-based training on objective performance measures and subjective evaluations in manual assembly tasks: A scoping review. *Human Factors*, *66*(2), 589–626.

DAQRI. (2015). 4 ways DAQRI Smart Helmet™ will transform the way you work. https://medium.com/@DAQRI_Media/4-ways-daqri-smart-helmet-will-transform-the-way-you-work-e5be12e164ee

DAQRI. (2016a, November 15). *Building the Worksite of the Future*. https://medium.com/%40DAQRI_Media/building-the-worksite-of-the-future-8326b0233006

DAQRI. (2016b). Building the worksite of the future. *Medium*. https://medium.com/%40DAQRI_Media/building-the-worksite-of-the-future-8326b0233006

DHL Customer Solutions & Innovation. (2023). *AI-Driven Computer Vision: A DHL Perspective on the Impact of Computer Vision in Logistics*. DHL Trend Research.

DHL Trend Research. (2014). *Augmented Reality in Logistics: Changing the Way We See Logistics – A DHL Perspective*. DHL Customer Solutions & Innovation.

Di Valentina, P., Valentina, D. S., Salvatore, M., & Stefano, R. (2021). Smart operators: How Industry 4.0 is affecting the worker's performance in manufacturing contexts. *Procedia Computer Science*, *180*, 958–967.

Domgue Kamwa, L. I., Paes, D., Feng, Z., Mander, S., Datoussaid, S., Descamps, T., et al. (2024). *Video See-Through Augmented Reality Fire Safety Training: A Comparison with Virtual Reality and Video Training*: Elsevier BV.

Doornbos, M.-C. J., Peek, J. J., Maat, A. P. W. M., Ruurda, J. P., Backer, P. de, Cornelissen, B. M. W., et al. (2024). Augmented reality implementation in minimally invasive surgery for future application in pulmonary surgery: A systematic review. *Surgical Innovation, 31*(6), 646–658.

Egger, J., & Masood, T. (2020). Augmented reality in support of intelligent manufacturing – A systematic literature review. *Computers & Industrial Engineering, 140*, 106195.

Emergen Research. (2025, October 3). *AR & VR in Training Market – Global Industry Report 2025*. Emergen Research. https://www.emergenresearch.com/industry-report/ar-vr-in-training-market

Eswaran, M., & Bahubalendruni, M.R. (2023). Augmented reality aided object mapping for worker assistance/training in an industrial assembly context: Exploration of affordance with existing guidance techniques. *Computers & Industrial Engineering, 185*, 109663.

Fernández-Moyano, J. A., Remolar, I., & Gómez-Cambronero, Á. (2025). Augmented reality's impact in industry – A scoping review. *Applied Sciences, 15*(5), 2415. https://www.mdpi.com/2076-3417/15/5/2415

Flor-Unda, O., Fuentes, M., Dávila, D., Rivera, M., Llano, G., Izurieta, C., & Acosta-Vargas, P. (2023). Innovative technologies for occupational health and safety: A scoping review. *Safety, 9*(2), 35. https://www.mdpi.com/2313-576x/9/2/35

Future Market Insights. (2025, October 7). AR and VR in Training Market Size and Share Forecast Outlook 2025–2035. *Future Market Insights*. https://www.futuremarketinsights.com/reports/ar-and-vr-in-training-market

Garcia, C., Ortega, M., Ivorra, E., Contero, M., Mora, P., & Alcañiz, M. L. (2024). Holorailway: An augmented reality system to support assembly operations in the railway industry. *Advances in Manufacturing, 12*(4), 764–783. https://doi.org/10.1007/s40436-023-00479-5

Gong, P., Lu, Y., Lovreglio, R., Lv, X., & Chi, Z. (2024). Applications and effectiveness of augmented reality in safety training: A systematic literature review and meta-analysis. *Safety Science, 178*, 106624.

Gualtieri, L., Revolti, A., & Dallasega, P. (2023). A human-centered conceptual model for integrating augmented reality and dynamic digital models to reduce occupational risks in industrial contexts. *Procedia Computer Science, 217*, 765–773.

Hancko, D., Majlingova, A., & Kačíková, D. (2025). Integrating virtual reality, augmented reality, mixed reality, extended reality, and simulation-based systems into fire and rescue service training: Current practices and future directions. *Fire, 8*(6), 228. https://www.mdpi.com/2571-6255/8/6/228

Hasan, I., Dhawan, P., Rizvi, S. A. M., & Dhir, S. (2023). Data analytics and knowledge management approach for COVID-19 prediction and control. *International Journal of Information Technology, 15*(2), 937–954. https://doi.org/10.1007/s41870-022-00967-0

Henderson, S., & Feiner, S. (2011). Exploring the benefits of augmented reality documentation for maintenance and repair. *IEEE Transactions on Visualization and Computer Graphics, 17*(10), 1355–1368.

Hussain, M., Ali, N., & Hong, J.-E. (2022). Vision beyond the field-of-view: A collaborative perception system to improve safety of intelligent cyber-physical systems. *Sensors, 22*(17), 6610. https://www.mdpi.com/1424-8220/22/17/6610

Ismael, M., McCall, R., McGee, F., Belkacem, I., Stefas, M., Baixauli, J., & Arl, D. (2024). Acceptance of augmented reality for laboratory safety training: Methodology and an evaluation study. *Frontiers in Virtual Reality, 5*, 1322543. https://doi.org/10.3389/frvir.2024.1322543

Jamson, A. H., Lai, F. C., & Carsten, O. M. (2008). Potential benefits of an adaptive forward collision warning system. *Transportation Research Part C: Emerging Technologies*, *16*(4), 471–484.

Kang, Y., Bowers, A. R., & Xu, J. (2025). Are directional or non-directional hazard warnings more helpful for drivers with homonymous visual field loss? *Applied Ergonomics*, *129*, 104615.

Khamaisi, R. K., Peruzzini, M., Raffaeli, R., Mangia, G., & Pellicciari, M. (2025). Designing adaptive human-cyber-physical systems to improve industrial tasks: A methodological approach. *The International Journal of Advanced Manufacturing Technology*, *136*(9), 4067–4082. https://doi.org/10.1007/s00170-024-14698-y

Kostadimas, D., Kasapakis, V., & Kotis, K. (2025). A systematic review on the combination of VR, IoT and AI technologies, and their integration in applications. *Future Internet*, *17*(4), 163. https://www.mdpi.com/1999-5903/17/4/163

Lampropoulos, G., Fernández-Arias, P., Antón-Sancho, Á., & Vergara, D. (2024). Examining the role of augmented reality and virtual reality in safety training. *Electronics*, *13*(19), 3952. https://www.mdpi.com/2079-9292/13/19/3952

Levy, N. (2016). How construction companies are using virtual reality to plan, build and manage complicated projects. *GeekWire*. https://www.geekwire.com/2016/construction-companies-turn-to-virtual-reality-to-plan-build-and-manage-complicated-projects/

Li, J., & Pilz, M. (2023). In-company training in a safety-critical industry: Lessons from the aircraft industry. *Journal of Workplace Learning*, *35*(2), 210–227. https://doi.org/10.1108/jwl-06-2022-0067

Li, X., Yi, W., Chi, H.-L., Wang, X., & Chan, A. P. (2018). A critical review of virtual and augmented reality (VR/AR) applications in construction safety. *Automation in Construction*, *86*, 150–162.

Liberatore, M. J., & Wagner, W. P. (2021). Virtual, mixed, and augmented reality: A systematic review for immersive systems research. *Virtual Reality*, *25*(3), 773–799. https://doi.org/10.1007/s10055-020-00492-0

Ma, J., Liu, Q., Yu, S., Li, X., & Liu, J. (2025). Prompting scientific concept learning in augmented reality: Impact of the concept map strategy on mental models and cognitive load. *Educational Technology Research and Development*, *73*(3), 1767–1790. https://doi.org/10.1007/s11423-025-10470-5

Maida, J. C., Bowen, C. K., & Pace, J. (2007). Improving robotic operator performance using augmented reality. *Proceedings of the Human Factors and Ergonomics Society Annual Meeting*, *51*(27), 1635–1639.

Mallam, S. C., Nazir, S., & Renganayagalu, S. K. (2019). Rethinking maritime education, training, and operations in the digital era: Applications for emerging immersive technologies. *Journal of Marine Science and Engineering*, *7*(12), 428. https://www.mdpi.com/2077-1312/7/12/428

Malta, A., Farinha, T., & Mendes, M. (2023). Augmented reality in maintenance-history and perspectives. *Journal of Imaging*, *9*(7), 142. https://www.mdpi.com/2313-433x/9/7/142

Mao, W., Scheffer, S., & Majumdar, A. (2025). Augmented reality-enabled knowledge management in industrial maintenance: The DILEAF framework. *Computers & Industrial Engineering*, *208*, 111363.

Mayáns-Martorell, J. (2013). Augmented user interface. *Procedia Computer Science*, *25*, 113–122.

Mehta, R. K. (2016). Integrating physical and cognitive ergonomics. *IIE Transactions on Occupational Ergonomics and Human Factors*, *4*(2–3), 83–87. https://doi.org/10.1080/21577323.2016.1207475

Moghaddam, M., & Klumpp, M. (2025). Conceptualizing cognitive and physical worker-technology symbiosis in manufacturing: Lessons from J.A.R.V.I.S. *Procedia Computer Science*, *253*, 661–672.

Morales Méndez, G., & Del Cerro Velázquez, F. (2024a). Augmented reality in Industry 4.0 assistance and training areas: A systematic literature review and bibliometric analysis. *Electronics*, *13*(6), 1147. https://www.mdpi.com/2079-9292/13/6/1147

Morales Méndez, G., & Del Cerro Velázquez, F. (2024b). Impact of augmented reality on assistance and training in Industry 4.0: Qualitative evaluation and meta-analysis. *Applied Sciences*, *14*(11), 4564. https://www.mdpi.com/2076-3417/14/11/4564

Nassereddine, H., Hanna, A. S., Veeramani, D., & Lotfallah, W. (2022). Augmented reality in the construction industry: Use-cases, benefits, obstacles, and future trends. *Frontiers in Built Environment*, *8*, 730094. https://doi.org/10.3389/fbuil.2022.730094

Palmarini, R., Del Amo, I. F., Ariansyah, D., Erkoyuncu, J. A., & Roy, R. (2023). Augmented reality for remote assistance (ARRA). In *Springer Handbook of Augmented Reality*, 669–685. https://doi.org/10.1007/978-3-030-67822-7_27

Palmarini, R., Erkoyuncu, J. A., Roy, R., & Torabmostaedi, H. (2018). A systematic review of augmented reality applications in maintenance. *Robotics and Computer-Integrated Manufacturing*, *49*, 215–228.

Park, K.-B., Kim, M., Choi, S. H., & Lee, J. Y. (2020). Deep learning-based smart task assistance in wearable augmented reality. *Robotics and Computer-Integrated Manufacturing*, *63*, 101887.

Penczek, J. (2025). Visual performance standards for virtual and augmented reality. *Frontiers in Virtual Reality*, *6*, 1575870.

PTC. (2023). *Vuforia Insights: Remote Expert Assistance and ROI Analysis*. PTC Inc.

Qeshmy, D. E., Makdisi, J., Da Ribeiro Silva, E. H. D., & Angelis, J. (2019). Managing human errors: Augmented reality systems as a tool in the quality journey. *Procedia Manufacturing*, *28*, 24–30.

Quandt, M., & Freitag, M. (2021). A systematic review of user acceptance in industrial augmented reality. *Frontiers in Education*, *6*, 700760. https://doi.org/10.3389/feduc.2021.700760

Rabhi, F. A., Bandara, M., Lu, K., & Dewan, S. (2021). Design of an innovative IT platform for analytics knowledge management. *Future Generation Computer Systems*, *116*, 209–219.

Rana, S., & Chicone, R. (2025). AI-enhanced virtual and augmented reality for cybersecurity training. In *Fortifying the Future* (pp. 101–131). Springer.

Rao, A. S., Radanovic, M., Liu, Y., Hu, S., Fang, Y., Khoshelham, K., Palaniswami, M., & Ngo, T. (2022). Real-time monitoring of construction sites: Sensors, methods, and applications. *Automation in Construction*, *136*, 104099. https://doi.org/10.1016/j.autcon.2021.104099

Rauschnabel, P. A., Felix, R., Hinsch, C., Shahab, H., & Alt, F. (2022). What is XR? Towards a framework for augmented and virtual reality. *Computers in Human Behavior*, *133*, 107289.

Rosak-Szyrocka, J. (2025). Reskilling and upskilling in the age of AI: A practical guide to workforce transformation. In *Artificial Intelligence, Machine Learning, Data Analytics and Automation for Business Management Series* (1st ed.). CRC Press LLC.

Rosak-Szyrocka, J., Gupta, S. K., & Boccia, F. (2024). New role of value for Customer 5.0 in augmented era. In *Aspects of Quality Management in Value Creating in the Industry 5.0 Way* (pp. 1–20). CRC Press.

Rosak-Szyrocka, J., Tripathi, S., Garcia, M. B., Festa, G., & Launer, M. A. (2025). *Reskilling and Upskilling in the Age of AI: A Practical Guide to Workforce Transformation*: CRC Press.

Rosak-Szyrocka, J., & Wolniak, R. (2026). *AI-driven sustainability: The future of human resources management*. CRC Press. https://doi.org/10.1201/9781003668282

Rosak-Szyrocka, J., Zywiołek, J., & Shahbaz, M. (2023). *Quality Management, Value Creation, and the Digital Economy* (1st ed.). Routledge.

Rosati, R., Romeo, L., Cecchini, G., Tonetto, F., Viti, P., Mancini, A., & Frontoni, E. (2023). From knowledge-based to big data analytic model: A novel IoT and machine learning based decision support system for predictive maintenance in Industry 4.0. *Journal of Intelligent Manufacturing*, *34*(1), 107–121. https://doi.org/10.1007/s10845-022-01960-x

Rymarov, A., Chandramauli, A., Sharma, G., Sharma, K., & Kumar, Y. (2024). Augmented reality and AI: An experimental study of worker productivity enhancement. *BIO Web of Conferences*, *86*, 1095.

Schwarz, F., & Fastenmeier, W. (2017). Augmented reality warnings in vehicles: Effects of modality and specificity on effectiveness. *Accident; Analysis and Prevention*, *101*, 55–66.

Scorgie, D., Feng, Z., Paes, D., Parisi, F., Yiu, T. W., & Lovreglio, R. (2024). Virtual reality for safety training: A systematic literature review and meta-analysis. *Safety Science*, *171*, 106372.

Söbke, H., Friedel, T., Wehking, F., Haupt, T., & Wiggenbrock, J. (2025). Workplace safety training using visual-simultaneous-localization-and-mapping-based mobile augmented reality. *Digital*, *5*(1), 8. https://www.mdpi.com/2673-6470/5/1/8

Solomashenko, A. B., Afanaseva, O. L., Shishova, M. V., Gulianskii, I. E., Sobolnikov, S. A., & Petrov, N. V. (2025). Industrial applications of AR headsets: A review of the devices and experience. *Light: Advanced Manufacturing*, *6*(2), 358. https://doi.org/10.37188/lam.2025.023

Song, Y., Koeck, R., & Luo, S. (2021). Review and analysis of augmented reality (AR) literature for digital fabrication in architecture. *Automation in Construction*, *128*, 103762.

Stefan, H., Mortimer, M., & Horan, B. (2023). Evaluating the effectiveness of virtual reality for safety-relevant training: A systematic review. *Virtual Reality*, *27*(4), 2839–2869. https://doi.org/10.1007/s10055-023-00843-7

Syed, T. A., Siddiqui, M. S., Abdullah, H. B., Jan, S., Namoun, A., Alzahrani, A., et al. (2022). In-depth review of augmented reality: Tracking technologies, development tools, AR displays, collaborative AR, and security concerns. *Sensors*, *23*(1), 146. https://www.mdpi.com/1424-8220/23/1/146

Tang, A., Owen, C., Biocca, F., & Mou, W. (2002). Experimental evaluation of augmented reality in object assembly task. In *Proceedings. International symposium on mixed and augmented reality* (pp. 265–266). IEEE.

Tatić, D., & Tešić, B. (2017). The application of augmented reality technologies for the improvement of occupational safety in an industrial environment. *Computers in Industry*, *85*, 1–10.

The Boeing Company. (2022). *Innovation Quarterly*, *6*(21). https://www.boeing.com

Thorslund, B., & Strand, N. (2016). Vision measurability and its impact on safe driving: A literature review. *Scandinavian Journal of Optometry and Visual Science*, *9*(1), 1–9. https://www.diva-portal.org/smash/record.jsf?pid=diva2:954681

Wang, S., Parsons, M., Stone-McLean, J., Rogers, P., Boyd, S., Hoover, K., et al. (2017). Augmented reality as a telemedicine platform for remote procedural training. *Sensors*, *17*(10), 2294. https://www.mdpi.com/1424-8220/17/10/2294

Winkler, M., Gabbard, J. L., Langlois, S., Kim, H., Nesti, A., Young, K. L., Merenda, C., & Nadri, C. (2025). A review of augmented reality heads up display in vehicles and human factors considerations. *International Journal of Human–Computer Interaction*. https://doi.org/10.1080/10447318.2024.2443252

Wolf, J., Wolfer, V., Halbe, M., Maisano, F., Lohmeyer, Q., & Meboldt, M. (2021). Comparing the effectiveness of augmented reality-based and conventional instructions during single ECMO cannulation training. *International Journal of Computer Assisted Radiology and Surgery*, *16*(7), 1171–1180. https://doi.org/10.1007/s11548-021-02408-y

Wu, S., Hou, L., Chen, H., Zhang, G., Zou, Y., & Tushar, Q. (2023). Cognitive ergonomics-based augmented reality application for construction performance. *Automation in Construction*, *149*, 104802.

Xia, L., Lu, J., Lu, Y., Zhang, H., Fan, Y., & Zhang, Z. (2024). Augmented reality and indoor positioning based mobile production monitoring system to support workers with human-in-the-loop. *Robotics and Computer-Integrated Manufacturing*, *86*, 102664.

Xie, J., Liu, S., & Wang, X. (2022). Framework for a closed-loop cooperative human Cyber-Physical System for the mining industry driven by VR and AR: MHCPS. *Computers & Industrial Engineering*, *168*, 108050.

Yazdi, M. (2024). Augmented reality (AR) and virtual reality (VR) in maintenance training. In *Advances in Computational Mathematics for Industrial System Reliability and Maintainability* (pp 169–183). https://doi.org/10.1007/978-3-031-53514-7_10

Yin, J.-H., Chng, C.-B., Wong, P.-M., Ho, N., Chua, M., & Chui, C.-K. (2020). VR and AR in human performance research——An NUS experience. *Virtual Reality & Intelligent Hardware*, *2*(5), 381–393.

You, Z., & Feng, L. (2020). Integration of Industry 4.0 related technologies in construction industry: A framework of cyber-physical system. *IEEE Access*, *8*, 122908–122922.

Yu, M., Lin, Q., & Lv, G. (2024). Safety augmentation for volitional human locomotion via lower-limb exoskeletons: A case study. *IEEE Control Systems Letters*, *8*, 778–783.

Zheng, N.-N., Tang, S., Cheng, H., Li, Q., Lai, G., & Wang, F.-W. (2004). Toward intelligent driver-assistance and safety warning system. *IEEE Intelligent Systems*, *19*(2), 8–11.

Zong, Z., & Guan, Y. (2025). AI-driven intelligent data analytics and predictive analysis in Industry 4.0: Transforming knowledge, innovation, and efficiency. *Journal of the Knowledge Economy*, *16*(1), 864–903. https://doi.org/10.1007/s13132-024-02001-z

Zywiołek, J., Rosak-Szyrocka, J., Nayyar, A., & Naved, M. (2024). *Modern Technologies and Tools Supporting the Development of Industry 5.0*. CRC Press.

6 AI-Powered Exoskeletons and Ergonomic Support

6.1 BIOMECHANICAL MONITORING AND DYNAMIC SUPPORT

The most significant evolutions in industrial ergonomics relate to human-machine collaboration with biomechanical monitoring by means of an artificial intelligence (AI)-powered exoskeleton. Conventional ergonomic systems depend either on static mechanical support or on generic kinematic assumptions, which cannot be tailored to fit the physiological character of workers or their fatigue level. An AI-powered exoskeleton fitted with sensors such as IMUs, electromyography (EMG) electrodes, and force-torque sensors will steadily acquire, interpret, and analyse biomechanical data for the context-sensitive delivery of mechanical assistive functions. This allows, much sooner than ever, the prediction of muscular strain, joint torque, and postural instability well in advance before all these factors reach their critical value. The exoskeleton is not a passive follower of human movements but rather an active cooperating agent for biomechanical optimization, guaranteeing an intelligent balance between physical efforts and mechanical assistance.

It would be for continuous biomechanical monitoring, whereby high-frequency streams coming from different nodes around the body, for example, shoulders, spine, hips, and knees, feed into an edge-based AI controller through recursive predictive modelling, generally via RNNs or adaptive Bayesian filters to denote asymmetry in motion, abnormal acceleration patterns, or changes in load distribution. These data will be correlated with an ergonomic risk model in which impending overload conditions are forecasted; dynamic adaptation is thus triggered by actuators via torque or resistance modulation. A particular example would be repetitive overhead assembly with progressive fatigue of the deltoid or trapezius muscles detected, therefore redistributing loads through the mechanical joints of the exoskeleton with compensatory lifting force but maintaining natural postural control. Thus, it becomes the dynamic extension of the musculoskeletal system of the worker by effectively preserving the biomechanical integrity even under an extended workload (Huang et al., 2025; Khan et al., 2025; Wang et al., 2025b).

The precision of such monitoring, however, rests squarely on the sophistication of sensor fusion mechanisms. Taken individually, sensors such as gyroscopes or EMG electrodes can provide only limited information. AI models combine such inputs in a multi-modal way into a coherent biomechanical profile that integrates both kinematic and kinetic data. This hybrid fusion architecture merges CNNs to provide spatial analysis with temporal pattern recognition models such as LSTM networks. The exoskeleton is thus positioned to perceive human motion in a holistic manner and capture subtle signs of fatigue in the form of micro-tremors or asymmetric torque distribution that would go unnoticed by human supervisors. This capacity thus allows

 DOI: 10.1201/9781003754213-8

the system to make proactive ergonomic corrections instead of reactive interventions in industrial contexts, thereby drastically reducing cumulative stress injuries (Ding et al., 2025; Liu et al., 2025; Winkler et al., 2025).

AI-powered exoskeleton biomechanical monitoring of work changes work safety from static to dynamic from a human factors perspective. The usual classical structure of ergonomic programmes is based on the periodic assessment and manual adjustment of workstation parameters. On the other hand, the integration of exoskeletons with AI turns optimization continuous and personalized. The AI can adapt, through adaptive feedback loops, for example, to support intensity in real time to the actual physiological state of the worker. Other secondary indicators the AI model might consider in modulation of the assistance are heart rate variability, skin temperature, and EMG fatigue indices. If muscular fatigue has just started, it increases torque contribution; when recovery is sensed, then the assistance is gradually reduced so as not to lose muscle engagement and avoid dependency. These cyclic adaptations preserve muscular strength over time, hence shielding the worker from overuse syndromes (Kumar et al., 2025; Morau et al., 2025; Zhao et al., 2025).

Exoskeletons can provide dynamic support, recognizing the context of a task. Through machine learning algorithms trained on historical data from industrial tasks, the system identifies the exact nature of the tasks, such as lifting, holding, bending, or some overhead assembly motion, which enables the AI to trigger only those biomechanical profiles pre-optimized for each type of movement. This would mean that during an overhead assembly, the AI would energize the upper-limb actuators so as to compensate for shoulder elevation torque, while the hip and lumbar actuators provide vertical support for all forms of lifting operations. The ability for context awareness will help avert biomechanical conflict between human intent and machine assistance. Without such synchronies, an exoskeleton might create mechanical resistance against natural movement, hence increasing fatigue rather than reducing it. Thus, modern exoskeleton intelligence is not about raw mechanical power but rather about the ability to predict and complement human biomechanics (Tang et al., 2025; Wang et al., 2025a; Zhu & Hu, 2025).

The other key ingredient in dynamic support is predictive adaptation by digital twins. More sophisticated implementations actually create a biomechanical digital twin of the worker: a virtual avatar simulating posture, motion trajectories, and joint loading over scenarios. From sensor data through the operating AI model running predictive analytics on the risks of ergonomics, the digital twin refreshes in real time. If it predicts excessive torque on the lumbar spine or cumulative micro-strain in the shoulder joint, it will make changes in the support parameters or send an early-warning alert. Predictive ability thus turns biomechanical monitoring from descriptive analytics – what happens – to prescriptive analytics – what needs to be done. In an industrial setting, for instance, automotive assembly or aerospace manufacturing, repetition over thousands of cycles makes anticipatory corrections absolutely necessary in avoiding chronic musculoskeletal disorders (Cetintav & Yalcin, 2025; Souaifi et al., 2025).

This is epitomized in the case of Boeing and EksoVest, whereby Boeing integrated passive exoskeletons into its assembly lines as a means of support for workers performing overhead tasks. The EksoVest, in itself, is not an active AI-driven system but catalyzes further development in the domain of intelligent exoskeletons. Measures of worker posture, range of motion, and fatigue, among other data that could be collected in pilot deployments, were useful in building AI models that recognize patterns associated with ergonomic risk. Such insights later inform hybrid designs marrying passive mechanical support with active AI monitoring. The reduction in shoulder strain by up to 30% and the decline in fatigue across its implementation at Boeing epitomize how biomechanical data can serve both as a diagnostic and optimization tool. It is experiences like this that accelerate the transition from static assistive devices to intelligent systems that learn from each motion cycle.

Besides performance enhancement, AI-driven biomechanical monitoring has huge implications for long-term health. MSDs are among the most common causes of occupational injury worldwide, with many leading to chronic disabilities and loss of productivity. Continuous monitoring of biomechanical load, keeping motions within safe thresholds, is a form of preventive healthcare integrated into workplaces through AI-powered exoskeletons. Over time, the system will build personalized biomechanical histories on each worker, possibly highlighting early degradations in joint stability or muscle performance. Data of this nature can be fed into occupational health programmes to enable predictive diagnostics and targeted physiotherapy long before any kind of injury. In that respect, the nexus of AI, ergonomics, and biomedical engineering ushers in a new paradigm of proactive human sustainability within industrial ecosystems (Dhaouadi et al., 2025; Jamshidian et al., 2025; Lefhal et al., 2026; Wang et al., 2026).

The three-tier architecture of the AI-powered exoskeleton includes a data acquisition layer acquiring raw sensor inputs from IMUs, EMG arrays, and pressure sensors; an intelligent processing layer where AI algorithms are used for feature extraction, pattern recognition, and decision optimization; and finally, the actuation layer, which converts these AI decisions into mechanical responses through torque modulation or changes in stiffness. The proposed closed-loop control structure will be able to respond within less than one second and maintain fluid motion. Local edge computing, supplying instant decisions with cloud-based analytics for long-term learning, provides responsiveness while continuously improving the models. Federated learning frameworks have recently seen higher frequencies in updating AI models from multiple users without compromising data privacy, a factor indispensable in industrial health monitoring (Dhaouadi et al., 2025; Laganà et al., 2025; Yan, 2025; Yang et al., 2025).

The ethical and organizational dimensions of biomechanical monitoring also cannot be ignored. On the contrary, AI-powered exoskeletons doubtless improve safety by reducing workforce fatigue, but they may generate a volume of biometric data. Governance mechanisms should be instituted so that such data gets used only for ergonomic and safety reasons, not spilling over into performance surveillance or control of the workforce. In this regard, transparent data policies and human-in-the-loop oversight are essential elements of any implementation strategy. User acceptance will also be dependent on psychological trust in the system's adaptability and nonintrusiveness. Training programmes that stress the symbiotic relation between human capability and machine intelligence help reduce resistance and engender a culture of collaborative safety (Cirillo et al., 2025; Kishor et al., 2025; Mittal et al., 2025).

Scientific and ethical dimensions are, therefore, at the core of biomechanical monitoring and dynamic support afforded by AI-powered exoskeleton systems. Connected with AI, such a system will give this support in a dynamic and learning-orientated way, upgrading ergonomic safety from static and mechanical to dynamic, predictive, and individual-centred. In such a way, it will not be only the present performance that will be preserved for the workers but, above all, their long-term physical integrity. This epitomizes the principles of human-centred automation within the wider framework of Industry 5.0, where intelligent technology amplifies human potential without sacrificing health or dignity (Abhinav et al., 2025; Gao et al., 2025; Kang, 2025; Mikołajewska et al., 2025).

6.2 TASK-SPECIFIC ADAPTATION AND USER FEEDBACK

The important divergences that new technology has made from classical ergonomics towards cognitive-adaptive human-machine interaction are the possibility of real-time adaptation of AI-powered exoskeletons to the particular context of tasks accomplished. In other words, whereas earlier generations of industrial exoskeletons depended on preconceived mechanical patterning logic, in a modern system, every task is modelled and classified; its functional properties are predicted with the help of AI. The capability for adaptation is also not confined to physical postures or load intensities but covers temporal, cognitive, and environmental aspects of work. It constantly monitors the current tasks being performed, tracks their development over time, and assesses how these factors modify the user's biomechanical and psychological condition. This will continually improve performance due to continuous learning loops so that in tempo and profile, the mechanical assist, the stiffness modulation, and energy use are matched with the demands imposed by specific operations, such as overhead fastening, lifting, welding, or assembling components in confined spaces (Kampouris & Azariadis, 2025; van der Kooij et al., 2025).

Context recognition inherently relies on multi-modal sensing and pattern classification. Onboard AI exploits continuous streams of sensor data – angular velocities, joint torques, muscle activation signatures, and even environmental ones like temperature or vibration – to identify the ongoing context of a task. These are possible because of convolutional and recurrent neural architectures that can segment user movements into micro-actions and then compare those against learned libraries of tasks. Those are generated through supervised and reinforcement learning on extensive datasets acquired from multiple workers performing standardized industrial activities. If there's a match, it will load a task profile onto the exoskeleton to define optimal torque distribution, actuator timing, and mechanical constraints. That same system is able to provide gentle, low-resistance support during precision tasks and strong, stabilizing torque during heavy or repetitive lifting without requiring manual reconfiguration of either.

Most salient with AI-based task adaptation is real-time elasticity, or in other words, prompt responsiveness to transitions or mixed operations. Rarely does industry have job cycles made of identical movements; instead, hybrid sequences including lifting, reaching, walking, and manipulation of tools are more common. Most classical exoskeletons consider this kind of heterogeneity a failure; they were designed for static assistance. Adaptive exoskeletons are in continuous reclassification of the flow of tasks and readjustment of mechanical response at millisecond latency. A typical example would be an aircraft component installation through AI recognition of transition from carrying into overhead fastening, redistributing the mechanical assistance from the lumbar module into the deltoid and triceps support modules. Seamlessly, without overassistance or mechanical lag interfering with the natural pattern of movement, the interaction pattern feels organic; with precise modulation of support, situational awareness may be maintained with fine motor control.

Personalized calibration enables adaptation of the task, considering users' differences. Such variables will be limb length, muscle strength, fatigue rate, and style of movement that will be part of a personal biomechanical profile recorded and tuned in the long term by an initial calibration session. An exoskeleton would compare in real time the motion signature of the worker with baselines of records while interacting with it and point to deviations from the nominal, hence signalling strain, stress, or inefficient posture. Machine learning models controlling parameters – often Gaussian process regressors or adaptive neuro-fuzzy inference systems – update to provide more support for a weaker user or reduce torque when too much assistance hampers agility. As pointed out by Rojek et al. (2023a) and Shah et al. (2023), connecting standardized ergonomic guidelines through great human morphological and work variety could thus be done by means of improved comfort, reduced energy expenditure, and superior user confidence thanks to individual adaptation by design.

The operation of the exoskeleton is, therefore, essentially based on a user feedback loop. AI-based exoskeletons themselves are not active machines operating on the body, part of a bidirectional flow of information with the human operator through several feedback channels: tactile, auditory, and optical. These means are designed to indicate system state, workload balance, and safety notifications. Modules of vibrotactile feedback – for example, integrated into shoulder straps or lumbar braces – can issue slight vibrations in case the posture of the user drifts away from an ergonomic optimum. Management of visual feedback via colour-coded indications of muscle load or system torque levels is projected onto smart glasses or wrist displays. Advanced models integrate voice or gesture recognition, thus enabling users to switch between modes of assistance without having to touch any controls. It is through such feedback systems that the worker maintains control of the degree of mechanical interference, thus preserving a sense of agency and avoiding the psychological discomfort associated with over-automation.

For many years, reinforcement learning has provided the framework within which user feedback is integrated into task adaptation in most industrial trials. Every user-exoskeleton interaction is considered by AI a reward-based learning episode in optimizing its support policy for maximum comfort and task performance. Thus, in cases where it often reduces assistance during certain motions, the system will infer that overassistance was occurring and lower the torque thresholds in future cycles. If the user continually signals fatigue or discomfort, then the system should raise the level of assistance or otherwise alter the pattern of motion synchrony. This in turn – after some time – will yield a stable, personalized control strategy. Such adaptive behaviour is highly beneficial at assembly lines or during logistics since tasks repeat daily while environmental conditions such as temperature, lighting, and vibration – as well as states of the humans concerning fatigue, hydration, and stress – change considerably (Kanakis et al., 2024; Tu & Qian, 2024).

The next step in this evolution is the integration of cognitive feedback. Cognitive feedback means that the system interprets not just biomechanical but also attentional and emotional indicators of user performance. AI algorithms already make it possible to infer states of stress, cognitive overload, or diminished concentration from physiological correlates such as heart rate variability, skin conductance, or microfacial expressions. For those places where these states are identified, the exoskeleton modulates behaviour accordingly – for example, by slowing the rate of movement or stabilizing balance – through auditory cues to take microbreaks. The holistic adaptation turns the exoskeleton from a purely mechanical assistant into a cognitive-ergonomic partner that protects physical and mental well-being alike. This feedback from the two areas reduces the possibility of error and improves sustained operational performance within safety-critical sectors such as aerospace, shipbuilding, and heavy logistics.

Another type of task-specific adaptation is environmental coadaptation, where the AI integrates contextual data from the workspace. The exoskeleton, linked into industrial Internet of Things systems, will receive environmental parameters such as ambient temperature, air quality, or even the presence of hazardous materials. This could enable the AI to adapt thermal management, ventilation, or actuator stiffness according to environmental stressors. Other examples include prioritizing cooling cycles when hot to avoid heat fatigue or initiating additional haptic guidance in areas with low visibility. Active digital twins of the production environment let exoskeletons simulate forthcoming task sequences and pre-align mechanical assistance for changes in conditions. Predictive alignment between user, task, and environment speaks to the intersection of cyber-physical ergonomics with AI-driven situational intelligence.

Another important dimension of adaptive systems is based on collaboration between users and tasks. Federated learning frameworks enable performance data from diverse exoskeleton units across a factory or building site to be shared, in anonymized form, towards the global improvement of the task recognition and assistance algorithms. In this sense, distributed intelligence will learn from a wide range of task configurations, materials, and ergonomic challenges without breaching personal privacy. If many workers go through similar strain patterns for some newly developed assembly process, for example, the global model would be able to infer an optimal support configuration and spread it over all devices. This would continuously enhance the efficiency of task adaptation at organizational levels, fostering an intelligent, connected workforce where human and machine are co-evolving.

Ethics regarding user acceptance and feedback need to be further developed. Task-adaptive exoskeletons collect sensitive information on the efficiency of motion, fatigue, and behavioural patterns that are at risk of being misused if not treated appropriately in monitoring or performance surveillance. Feedback loops must be designed to be transparent, with the user clearly in control over what data is shared and how the levels of assistance are set to avoid distrust. Participatory design methods, where workers themselves can help in defining the parameters of feedback and adaptation, have proved to offer the best way of finding a balance between ethical considerations and the assurance of higher system adoption rates. The trust in the adaptive system follows through with the user's perception of it as augmenting, not replacing, human autonomy. Feedback should focus on underlining collaboration, self-awareness, and personal safety rather than evaluation or productivity metrics.

These are challenging technical requirements, since in particular task recognition models must balance the trade-off between precision and computational efficiency, given that most of the exoskeletons rely on edge computing with limited energy capacity. Basic recognition algorithms, including lightweight temporal convolutional networks or even spiking neural networks, can yield good recognition accuracy with low latency at low power consumption. User feedback requires richer, more robust signal-processing pipelines that can overcome noise from sensor drift, motion artefact, and environmental interference. Hybrid control schemes combine model-based dynamic control with AI-driven learning and show the best features in terms of stability and responsiveness for research prototypes. Such architectures guarantee adaptation to the task in an interpretable way, meaning that it is possible for engineers and occupational specialists to audit the logic of the decisions and check whether the standard operation is performed in compliance with safety standards (Wang et al. 2025c; Zhu & Hu, 2025).

Thus, put in perspective, synergy between task-specific adaptation and continuous feedback heralds a new paradigm: 'ergonomic intelligence.' Instead of fixed mechanical parameters, the system transforms into a living interface that continuously learns from human experiences and contextual variation. This change falls in line with wider principles of Industry 5.0, emphasizing such aspects as human-centricity of technology, personalization, and resilience. Adaptive exoskeletons epitomize how AI, cyber-physical sensing, and human cognitive feedback have merged into an industrial environment wherein, in the future, safety, efficiency, and dignity shall coexist. Moving forward, their role shall extend from injury prevention to performance augmentation, supporting rehabilitation, and cultivating sustainable, health-conscious industrial ecosystems beyond what (Lee et al., 2024; Tu & Qian, 2024) have reported.

Table 6.1, Operational framework of task-specific adaptation and user feedback mechanisms in AI-powered exoskeletons. It depicts how biomechanical intelligence interacts with contextual awareness and artificial learning. Each row represents one critical dimension of adaptive ergonomics, ranging from contextual recognition and personalized calibration to predictive task transition and ethical data governance. These will collectively uncover how the exoskeleton will evolve from static assistive devices into an intelligent, co-adaptive system that learns from human motion, environmental cues, and physiological states. Techniques like convolutional and recurrent neural networks, reinforcement and federated learning, or neuro-fuzzy inference systems can be embedded to enable a continuous optimization process where devices perceive events, interpret them, and change their behaviour in real time. This will guarantee symbiotic interaction between human and machine, dynamic personalization of ergonomic support, proactive maintenance of safety, and continued centrality of human autonomy within technological augmentation.

TABLE 6.1
Operational Framework of Task-Specific Adaptation and User Feedback Mechanisms in AI-Powered Exoskeletons

Aspect	Description of Task-Specific Adaptation/Feedback Mechanism	AI Techniques and Implementation Examples
Context recognition	The system identifies the current task in dependence on motion, load, and environmental parameters. Using multi-modal sensor inputs, it will classify if it is lifting, overhead assembly, precision handling, or walking tasks.	Motion sequences are classified by Convolutional Neural Networks and Recurrent Neural Networks in real time, thus triggering task-specific profiles.
Dynamic recalibration	It automatically changes torque, stiffness, and level of support when a worker changes tasks or posture.	Adaptive control with temporal pattern recognition and continuous retraining of the model with reinforcement learning.
Individualized calibration	A personalized biomechanical profile is created for each user, accounting for anthropometry, muscle capacity, and fatigue dynamics.	Gaussian Process Regression and Neuro-Fuzzy Inference Systems allow for real-time adjustments concerning user-specific biomechanics.
Real-time elasticity	The system allows immediate switching between task modes without interrupting the movement, hence assuring smooth cooperation with the user.	Adaptive controllers at the edge and predictive Bayesian filters reduce latency in movement transitions.
User feedback loop	Continuous information of the load balance and posture is conveyed back to the user through tactile, auditory, or visual feedback.	Vibrotactile actuators, augmented reality visual cues, and voice-command interfaces for two-way interaction.
Reinforcement learning from feedback	It fine-tunes its assistant policy, collecting user satisfaction, comfort, and correction signals during work sessions.	The reinforcement learning (RL) models optimize torque thresholds and motion synchronization patterns based on user responses.
Cognitive feedback	It interprets physiological and psychological indicators of stress or the loss of concentration to modify behaviour.	Biosignal processing and emotion-recognizing AI can detect anomalies in heart rate variability, skin conductance, and facial microexpressions.
Environmental coadaptation	The integration with Industrial Internet of Things (IIoT) sensors will further let the exoskeleton adapt its support based on ambient conditions, like temperature or air quality.	Context-aware AI fuses IoT data streams for dynamic adjustments in ventilation, thermal management, or actuator stiffness.
Collaborative learning across users	The data from anonymous tasks across many exoskeletons are pooled, improving model robustness through increased generalization of task recognition.	Federated learning frameworks update the shared models without transferring personal data, hence guaranteeing privacy and scalability.

(Continued)

TABLE 6.1 (CONTINUED)

Aspect	Description of Task-Specific Adaptation/Feedback Mechanism	AI Techniques and Implementation Examples
Predictive task transition	The AI predicts, based on the history of the user and supported by digital twin simulations, which movements or tasks will likely occur next.	Predictive modelling and time-series forecasting using LSTM architectures for anticipatory torque planning.
Ethical feedback governance	Trust and acceptability are warranted by the user's control over data collection and the channels for feedback.	Transparent feedback management is defined by privacy-preserving protocols and participatory design principles.
Multi-modal interaction	The user can also interface with the system through his or her voice, gesture, or wearables to change parameters for assistance.	The natural language processing (NLP) and gesture-recognition models support intuitive command inputs.
Error detection and self-correction	AI automatically identifies inconsistencies between human intention and mechanical response, while self-calibration is achieved automatically.	In order to ensure the safety of ergonomics during fast motions, fault detection algorithms are performed in a hybrid model-based and data-driven way.
Performance optimization	Continuous feedback allows the system to learn the best assistance strategy, balancing effort, safety, and efficiency.	Multi-objective optimization algorithms encompass simultaneous energy efficiency, comfort, and task accuracy.
Human-machine symbiosis	The ultimate goal is a cooperative equilibrium at which human motion and machine support will merge.	Continuous reinforcement cycles and self-tuning AI controllers maintain alignment between human intent and mechanical actuation.

Source: Author's own work.

6.3 STRATEGIES FOR LONG-TERM HEALTH AND INJURY PREVENTION

For AI-powered exoskeletons to find long-term integration within industrial and healthcare environments, the focus needs to shift from optimization of short-term ergonomics towards sustained health management and preventive biomechanics. Whereas assistive devices of old met basic needs by reducing fatigue or simply stabilizing posture, current and future generations must be concerned with the far more ambitious pursuit of ensuring cumulative musculoskeletal health, regenerative support, and adaptive conditioning. AI, in this respect, is an active guardian-teacher: while closely monitoring subtle deviations of motion, it looks ahead to the prospect of injury over the next several years and guides the user towards biomechanical efficiency. The challenge is essentially based on a continuous feedback ecosystem in which human physiology, mechanical design, and AI-driven analytics continuously interact to maintain physical integrity over years of repetitive effort.

While the development of this system is still in progress, it could one day safeguard workers from back injuries. Core to long-term injury prevention, preventive biomechanics means AI-driven exoskeletons continuously monitor the mechanical load on critical anatomical structures such as the lumbar spine, shoulder complex, knees, and wrists, and ensure that movement patterns remain in physiologically safe ranges. Predictive models inside, informed by big data sets of human motion, detect patterns of cumulative micro-strain precursory to musculoskeletal disorders. For example, slight asymmetry during shoulder elevation, should it be repeated thousands of times a week, will not cause pain but will lead to chronic tendinopathy. It is here that the AI engine identifies micro-patterns through sophisticated feature extraction techniques and either warns the user or redistributes support automatically to counteract the imbalance. Over time, this kind of compensation becomes biomechanical prophylaxis, intelligent safeguards against repetitive strain injuries that are otherwise invisible in their early stages (Amin et al., 2025; Irfan et al., 2025; Kim et al., 2025).

Exoskeletons operationalize the concept of preventive biomechanics through predictive health analytics by integrating real-time biomechanical data with longitudinal performance trajectories. Machine learning algorithms – most notably, long short-term memory networks and gradient boosting models – analyse longitudinal trends in joint torque, muscular activation, and range-of-motion variability to produce personalized ‘risk profiles’ that quantify exposure to potential injury. For example, an algorithm may detect an increasing cumulative shoulder load index for a given user and recommend micro-rest intervals, ergonomic corrections, or changes in movement sequencing. In the more advanced systems, such predictive insights flow into occupational health platforms where medical staff can trace the evolution of individual risk over months or years. AI-powered exoskeletons bridge industrial safety engineering and personalized occupational medicine by converting biomechanical data into preventive healthcare indicators.

Adaptive rehabilitation is the nearest future application of AI-powered exoskeletons, considering not only injury prevention but also active recovery and regeneration. With the post-injury or post-surgery stage in mind, active AI-controller-powered exoskeletons can guide such a step-by-step process of reintroduction of loads to the healing muscle or joint, progressing them precisely to not overexert them. Examples of biomechanical and physiological markers that monitor the status of recovery include range of motion, torque symmetry, and EMG signal amplitude. Through adaptive algorithms, the exoskeleton adjusts the amount of assistance according to daily improvement metrics that allow for safe neuromuscular pathway reactivation. These replace static rehabilitation protocols with adaptive, data-driven therapies tuned to an individual’s trajectory of healing (Flaker et al., 2025; Saravanan et al., 2025; Wang et al., 2025a).

This therapeutic potential is furthered by neuromuscular feedback integration; AI algorithms interpret the EMG signals not only as a representation of effort but also as a stimulus for retraining neural coordination. Application can be made in the long-term health context: to counteract muscle atrophy, motor asymmetry, or neuromuscular inefficiency brought on by repetitive occupational patterns. Continuous calibration by the exoskeleton stimulates balanced muscular recruitment, prevents dominance of particular groups of muscles, and fosters postural equilibration. The result of this, over time, is musculoskeletal resilience – a physiological condition described by symmetry, flexibility, and endurance. An integration of AI-assisted rehabilitation with preventive ergonomics thus represents a seamless continuum of physical preservation (Butler-Henderson et al., 2025; Kajornkasirat et al., 2025; Yan et al., 2025).

Other important and long-term dimensions of health preservation include load management and micro-recovery cycles. Traditional ergonomics relies on some generalized time-motion studies to determine rest intervals, with little in the way of precision. Dynamic profiling of loads using AI-based exoskeletons allows for suggestions or enforcement of microbreaks at times determined not by fixed schedules but by real-time fatigue indicators. For example, fatigue-detecting algorithms may make use of shifts in EMG frequency, local vibration amplitude, or lowered motion smoothness to infer muscular overload. Early signs of fatigue would thus trigger localized mechanical offloading or recommend brief rest or otherwise redistribute work intensity to less fatigued muscle groups.

It is further based on a physiological load cycling principle, mimicking the mechanisms of biological recovery by periodically providing support and partial load exposure. Optimal cycles, worked out by AI, keep the muscles active without straining and hence improve short-term performance and long-term tissue recovery. More than 25% reduction in cumulative fatigue indices due to predictive load management in large-scale automotive assembly and logistics extends the productive horizons of workers. In the future, it will be able to integrate metabolic states measured by wearable sensors – oxygen consumption, lactate concentration, or skin temperature – and allow exoskeletons to synchronize the support cycles with real metabolic recovery states and hence permit fully personalized fatigue management systems (Saravanan et al., 2025; Wang et al., 2025b; Wuryantoro & Sabah, 2025).

It is on this scale of organizational level that the long-term health strategy expands into the realm of ergonomic intelligence. AI-powered exoskeletons produce volumes of data from biomechanical stress patterns and environmental influences to user behaviour in the workforce. Aggregated together, these data paint a picture of structural inefficiencies either in workplace design, task sequencing, or equipment layout. For example, if aggregated exoskeleton data show repeated lumbar strain among workers working at a certain workstation, that could indicate the need for spatial redesign or task redistribution. This endless loop of ergonomic intelligence turns the workplace into a self-improving ecosystem rather than a static set of safety protocols (Belov et al., 2025; Hole et al., 2025; Prasad, 2025; Pratap & Arulkumaran, 2025; Vats et al., 2025).

Central to this transformation will be collective learning, with analytics driven by AI. Federated and swarm learning models process in a decentralized way the data of many exoskeletons. These approaches maintain privacy while creating global insights into ergonomics. The resultant patterns feed into machine learning algorithms and organizational decision-making. Over time, the system evolves into a knowledge infrastructure able to anticipate which tasks bear long-term risks and how those can be reduced by changes in design or workflow. In this manner, AI-powered exoskeletons will be supportive not only of injury prevention for individual workers but also for systemic redesign towards human-centred production environments.

Biomechanical optimization cannot preserve long-term health; psychophysiological and cognitive issues are involved as well. Continuous physical strain in conjunction with monotonous motion is closely related to cognitive fatigue, reduced attentional control, and psychosomatic tension. AI-powered exoskeletons will alleviate these effects through biosignal feedback mechanisms monitoring heart rate variability, skin conductance, or ocular micro-movements as biomarkers for stress. If, in this case, these biomarkers distinctly reflect a decrease in concentration or an increase in stress, the system would introduce compensatory action, such as adjustments in assistance, slowing down the pace of tasks, or auditory reminders for micro-rests and hydration (Mi & Huang, 2025; Reyes-Hernández et al., 2025; Shingne et al., 2025; Wuryantoro & Sabah, 2025).

Beyond monitoring, such systems can enhance cognitive sustainability by increasing users' awareness of body mechanics and mental state. Real-time AR visualizations of posture symmetry, joint load, or breathing rhythm enable users to develop a sense of ergonomic self-regulation. The biofeedback-based learning has long-term beneficial effects on body consciousness, one of the most important psychological factors for injury prevention and maintaining motivation. Biomechanical intelligence combined with affective computing extends exoskeleton functions into the cognitive sphere and creates coherence between physical support and mental toughness. (Hole et al., 2025; Zim & Hanif, 2025).

AI-powered exoskeleton systems have to be embedded into occupational health policy and a preventive health framework to create a long-term impact. If privacy and ethical standards are upheld, continuous data can be the basis for lifelong health tracking based on movement patterns, load history, and episodes of fatigue. In forward-looking industrial organizations, this may be used to feed data into their digital occupational health records in order to support early diagnostics of musculoskeletal decline and evidence-based rehabilitation planning. This fits within the greater paradigm of predictive and personalized medicine, whereby health management becomes proactive preservation rather than reactive treatment.

Successful implementation does, however, require an ethical and governance framework: transparency of data use, anonymization protocols, and active human oversight mechanisms working to preserve worker trust. The use of health-related data for punitive evaluation or performance ranking would have to be proscribed. Instead, organizational policies should classify exoskeleton data as part of a shared health resource, accessible by employee and employer in view of cooperative management of well-being. This co-governance model represents Industry 5.0 principles of human dignity, safety, and empowerment within a technologically augmented workplace (Mi & Huang, 2025; Shingne et al., 2025; Wang et al., 2025c).

Health, in a long-term sense, means an extension of the span of healthy, productive working life through technological symbiosis. From biomechanics to AI and physiological analytics, the exoskeleton redefines labour, health, and sustainability by letting workers maintain functional capacity beyond the norm due to ageing or accumulated fatigue.

This would involve devising adaptive ageing frameworks whereby AI-driven exoskeletons update their support profiles according to the evolution in the user's physiological characteristics over time. For instance, compensation for age-related losses in muscle elasticity or proprioception could be performed gradually with a view to sustaining performance and safety throughout the career span. Its relevance is related to a socio-technical evolution from reactive protection of the body to proactive cultivation of human vitality.

BIBLIOGRAPHY

Abhinav, V., Basu, P., Verma, S. S., Yadav, P. R., & Kumar, V. (2025). Advancements in wearable and implantable BioMEMS devices: Transforming healthcare through technology. *Micromachines*, *16*(5), 522.

Amin, P., Gulzar Ahmad, S., Khan, H. U., Munir, E. U., & Ramzan, N. (2025). Predictive analytics in maternal health: A machine learning approach for classification of preeclampsia. *International Journal of Computational Intelligence Systems*, *18*(1), 115.

Belov, D., Chen, W., Liland, E., & Malik, A. (2025). Assessing drill string component health through predictive digital analytics. In *SPE International Association of Drilling Contractors Drilling Conference Proceedings*, 2025 March.

Berdell, J., Kudernatsch, S., & Ferdowsi, H. (2024). AI-driven solid-state device to enable natural control of upper-extremity robotic exoskeletons. *Systems Science and Control Engineering*, *12*(1), 2331061.

Butler-Henderson, K., Arabi, S., & Wang, W. (2025). Data's big three: How learning health systems, artificial intelligence and predictive analytics are transforming healthcare. *BMJ Health & Care Informatics*, *32*(1), e101414.

Cetintav, B., & Yalcin, A. (2025). Decoding canine behaviors through explainable AI and wearable sensor data. *Journal of Veterinary Behavior*, *81*, 1–9.

Chang, J., Li, J., Ye, J., Tan, J.-C., & Zhao, H. (2025). AI-enabled piezoelectric wearable for joint torque monitoring. *Nano Micro Letters*, *17*(1), 247.

Cirillo, E., Conte, C., Moccardi, A., & Fonisto, M. (2025). Resilient AI in therapeutic rehabilitation: The integration of computer vision and deep learning for dynamic therapy adaptation. *Applied Sciences*, *15*(12), 6800.

Coser, O., Tamantini, C., Soda, P., & Zollo, L. (2024). AI-based methodologies for exoskeleton-assisted rehabilitation of the lower limb: A review. *Frontiers in Robotics and AI*, *11*, 1341580.

Dhaouadi, S., Khelifa, M. M. B., Balti, A., & Duché, P. (2025). Optical sensor-based approaches in obesity detection: A literature review of gait analysis, pose estimation, and human voxel modeling. *Sensors*, *25*(15), 4612.

Ding, Q., Li, R., Liu, Q., & Cui, W. (2025). Human-powered electricity generation: Current technologies, challenges, and potential application in sustainable society construction. *Energy Conversion and Management X*, *28*, 101239.

Flaker, S. M., Holm, M., Gilmer, M., Perry, A., & Martindale-Mathern, S. (2025). Optimizing influenza vaccine allocation: A predictive analytics approach for informed public health planning. *American Journal of Health-System Pharmacy*, *82*(10), 544–550.

Gao, Y., Xu, B., Qiu, M., Guan, X., & Fu, H. (2025). Fabric-reinforced functional insoles with superior durability and antifracture properties for energy harvesting and AI-empowered motion monitoring. *Advanced Functional Materials*, *35*(10), 2416577.

Hole, S. R., Bhavekar, G. S., Prajapati, A. K., Karpe, S. R., & Vrindavanam, J. (2025). Hybrid PCA-based machine learning models for predictive analytics in urban health monitoring systems. In *Proceedings of the IEEE 1st International Conference on Smart and Sustainable Developments in Electrical Engineering (SSDEE 2025)*.

Huang, Y., Antwi-Afari, M. F., Sun, B., & Liu, J. (2025). Critical success factors for implementing self-powered wearable internet of things sensors in construction: A systematic literature review and conceptual framework. *Applied Energy*, *401*, 126836.

Irfan, N., Zafar, S., Shakil, K. A., Jaiganesh, A., & Abubeker, K. M. (2025). Integrating AI predictive analytics with naturopathic and yoga-based interventions in a data-driven preventive model to improve maternal mental health and pregnancy outcomes. *Scientific Reports*, *15*(1), 23878.

Iyer, V., Dhivya, J., Rani, P. A. A., Harivignesh, C. K., & Mohamed Farhan, P. (2025). AI-driven motion analysis of lower limb exoskeleton with integrated backpack load. In *Proceedings of the 3rd International Conference on Advancements in Electrical, Electronics, Communication, Computing and Automation (ICAECA 2025).*

Jamshidian, M., Wittek, A., Sekhavat, S., Wood, C., & Miller, K. (2025). 4D-CTA image and geometry dataset for kinematic analysis of abdominal aortic aneurysms. *Data in Brief*, *61*, 111797.

Kajornkasirat, S., Sawangwong, C., Puangsuwan, K., Phutthamongkhon, W., & Puttinaovarat, S. (2025). Integrating AI-driven predictive analytics in wearable IoT for real-time health monitoring in smart healthcare systems. *Applied Sciences (Switzerland)*, *15*(8), 4400.

Kampouris, C., & Azariadis, P. (2025). A methodology for rapid deployment of diagnostic applications for human gait and posture analysis and exoskeleton configurations, combining AI with a biomimetic rig. *Artificial Intelligence and Applications*, *3*(1), 10–30.

Kanakis, A., Katsampiris-Salgado, K., Zacharaki, N., Dimitropoulos, N., & Makris, S. (2024). Cognitive exoskeletons: Harnessing AI for enhanced wearable robotics in shipbuilding. *Lecture Notes in Mechanical Engineering*, 126–135.

Kang, Y. J. (2025). Microfluidic chip for quantitatively assessing hemorheological parameters. *Micromachines*, *16*(5), 567.

Khan, I., Murad, S., Khalid, M., Ramzan, H., & Khan, H. H. (2025). Wearables, smart textiles & AI biomechanics in sports: A narrative review. *Premier Journal of Science*, *14*, 100112.

Kim, T.-H., Srinivasulu, A., Chinthaginjala, R., Obaidur Rab, S., & Tera, S. P. (2025). Improving CNN predictive accuracy in COVID-19 health analytics. *Scientific Reports*, *15*(1), 29864.

Kishor, I., Goyal, P., Reddy, H. G., Goyal, P., & Mamodiya, U. (2025). AI-driven haptic technologies revolutionizing patient rehabilitation. In *Integrating AI with Haptic Systems for Smarter Healthcare Solutions* (pp. 47–84). IGI Global.

Kumar, R., Waisberg, E., Ong, J., Zaman, N., & Tavakkoli, A. (2025). Precision health monitoring in spaceflight with integration of lower body negative pressure and advanced large language model artificial intelligence. *Life Sciences in Space Research*, *47*, 57–60.

Laganà, F., Pellicanò, D., Arruzzo, M., Pullano, S. A., & Fiorillo, A. S. (2025). FEM-based modelling and AI-enhanced monitoring system for upper limb rehabilitation. *Electronics*, *14*(11), 2268.

Lee, D., Lee, S., & Young, A. J. (2024). AI-driven universal lower-limb exoskeleton system for community ambulation. *Science Advances*, *10*(51), eadq0288.

Lefhal, D., Ouacha, A., El Harraj, A., & Ziti, S. (2026). Advancements in emerging technologies for athlete health monitoring, physical preparedness, and injury prevention in team sports: A comprehensive narrative review. *Lecture Notes in Networks and Systems*, *1586*, 308–319.

Liu, S., Duan, X., Shi, L., Wang, Y., & Peng, L. (2025). Bio-inspired fully flexible triboelectric nanogenerator with AI-assisted context-awareness for elderly fall detection and injury localization. *Applied Physics A: Materials Science and Processing*, *131*(9), 665.

Mi, Q., & Huang, C. (2025). Integrating predictive analytics and digital health innovations to reduce postoperative emergency department admissions in bariatric surgery. *Obesity Surgery*, *35*(3), 1193–1194.

Mikołajewska, E., Mikołajewski, D., Mikołajczyk, T., & Paczkowski, T. (2025). A breakthrough in producing personalized solutions for rehabilitation and physiotherapy thanks to the introduction of AI to additive manufacturing. *Applied Sciences*, *15*(4), 2219.

Misir, A. (2025). Current developments in orthopaedic implant technology. *Journal of Orthopaedic Surgery and Research*, *20*(1), 927.

Mittal, S., Chandel, A., & Ta, H. H. (2025). AI-based player fatigue and workload monitoring systems. In *AI and Machine Learning Applications in Sports Analytics* (pp. 15–39). IGI Global.

Morau, S., Macedo, L., Morais, E., Martinek, R., & Leal-Junior, A. (2025). Low-cost AI-enabled optoelectronic wearable for gait and breathing monitoring: Design, validation, and applications. *Biosensors*, *15*(9), 612.

Prasad, A. (2025). Predictive analytics in clinical psychology: Role of machine learning in future mental health care. In *Transforming Neuropsychology and Cognitive Psychology with AI and Machine Learning* (pp. 313–332). IGI Global.

Pratap, M. S., & Arulkumaran, G. (2025). Machine learning-based predictive analytics for health diagnosis: A case study on diabetes prediction. In *Proceedings of the 3rd International Conference on Inventive Computing and Informatics (ICICI 2025)* (pp. 500–505).

Reyes-Hernández, Y., Ruiz-Vanoye, J. A., Rodríguez-Flores, J., Marroquín-Gutiérrez, F., & Salgado-Ramírez, J. C. (2025). Predictive analytics and machine learning in public health. In *Machine and Deep Learning Solutions for Achieving the Sustainable Development Goals* (pp. 165–180). IGI Global.

Rojek, I., Dorożyński, J., Mikołajewski, D., & Kotlarz, P. (2023a). Overview of 3D printed exoskeleton materials and opportunities for their AI-based optimization. *Applied Sciences*, *13*(14), 8384.

Rojek, I., Mikołajewski, D., Dostatni, E., & Kopowski, J. (2023b). Specificity of 3D printing and AI-based optimization of medical devices using the example of a group of exoskeletons. *Applied Sciences*, *13*(2), 1060.

Rosak-Szyrocka, J., & Wolniak, R. (2025). *AI-Driven Sustainability: The Future of Human Resources Management* (p. 224). Taylor & Francis Group.

Saravanan, S. K., Parvathy, P. R., Meenakshi, R., Ganeshbabu, T. R., & Meenakshi, B. (2025). Predictive analytics for ear health management using IoT and ensemble learning. In *Proceedings of the International Conference on Intelligent Control Computing and Communications (IC3 2025)* (pp. 319–324).

Shah, J. M., Natraj, N. A., Hallur, G. G., & Aslekar, A. (2023). Artificial intelligence (AI) in the automotive industry and the use of exoskeletons in the manufacturing sector of the automotive industry. In *Proceedings of the 2nd International Conference on Sustainable Computing and Data Communication Systems (ICSCDS 2023)* (pp. 428–432).

Shingne, S., Bhosale, M., Gosavi, T., Meshram, M., & Navalkar, S. (2025). Comprehensive health forecasting and care solutions using predictive analytics and AI techniques. In *Proceedings of the International Conference on Emerging Trends in Engineering and Technology (ICETET 2025)*.

Skubis, I., Wolniak, R., & Grebski, W. W. (2024). AI and human-centric approach in smart cities management: Case studies from Silesian and Lesser Poland Voivodships. *Sustainability*, *16*, 18.

Souaifi, M., Dhahbi, W., Jebabli, N., Muntean, R. I., & Dergaa, I. (2025). Artificial intelligence in sports biomechanics: A scoping review on wearable technology, motion analysis, and injury prevention. *Bioengineering*, *12*(8), 887.

Tang, C., Yi, W., Zhang, Z., Occhipinti, E., & Occhipinti, L. G. (2025). AI-driven smart sportswear for real-time fitness monitoring using textile strain sensors. *IEEE Transactions on Biomedical Engineering*, *73*(2), 555–565.

Tu, C., & Qian, C. M. (2024). AI-enabled video diagnosis and assisted walking with exoskeleton for Parkinson's disease. *Lecture Notes in Electrical Engineering*, *1214*, 145–154.

van der Kooij, H., van Asseldonk, E. H. F., Sartori, M., Esser, A., & Riener, R. (2025). AI in therapeutic and assistive exoskeletons and exosuits: Influences on performance and autonomy. *Science Robotics*, *10*(104), eadt7329.

Vats, P., Kshirsagar, P. R., Upreti, K., Tak, T. K., & Mahajan, S. (2025). Behavioral analytics for predictive modeling of mental health disorders: A review of machine learning techniques and challenges. In *Proceedings of the International Conference on Intelligent Control Computing and Communications (IC3 2025)* (pp. 376–381).

Venkatesh, J., Deepa, A., Mohammed, U. U. M., Selvamanikandan, R., & Sarikha, V. K. (2025). Autonomous exoskeletons for real-time gait analysis and rehabilitation with AI for mobility-impaired patients. In *Proceedings of the 6th International Conference on Inventive Research in Computing Applications (ICIRCA 2025)* (pp. 1861–1866).

Vittoria Guerra, B. M., Sozzi, S., Pizzocaro, S., De Nunzio, A. M., & Schmid, M. (2024). AI processing of wearable IMU data for exoskeleton gait analysis. In *Proceedings of the 8th IEEE International Forum on Research and Technologies for Society and Industry Innovation (RTSI 2024)* (pp. 380–384).

Wang, H., Song, Y., & Bi, H. (2025a). Optimizing public health management with predictive analytics: Leveraging the power of random forest. *Frontiers in Big Data*, *8*, 1574683.

Wang, K., Cui, J., Li, X., Zheng, Y., & Xue, C. (2025b). A flexible MWNT/PVDF electrospun membrane-based triboelectric nanogenerator with a spatially ordered hollow array for musculoskeletal disorder monitoring. *Chemical Engineering Journal*, *523*, 168847.

Wang, L., Guo, X., Zhang, Z., & Lee, C. (2025c). Metaverse-enabled yoga coach avatar using AI-enhanced multimodal insole sensing system. *Advanced Functional Materials*, *36*(20), 2519562.

Wang, P., Wang, A., & Wang, S. (2026). Integrating multimodal AI technologies for sports injury prediction and rehabilitation: Systematic review. *Journal of Human Sport and Exercise*, *21*(1), 22–37.

Winkler, E. V., Lauer, S. K., & Steigmeier-Raith, S. (2025). Kinematic gait analysis in dogs: Is modern gait analysis finding its way into veterinary practice? *Tierärztliche Praxis Ausgabe K: Kleintiere Heimtiere*, *53*(5), 298–307.

Wolniak, R., & Stecuła, K. (2024). Artificial intelligence in smart cities: Applications, barriers, and future directions—A review. *Smart Cities*, *7*, 3.

Wuryantoro, A., & Sabah, H. A. (2025). Computational IoT framework with GPRS integration for real-time camel health monitoring, predictive intervention, and data analytics in desert ecosystems. *Lecture Notes in Networks and Systems*, *1305*, 185–194.

Yan, C. (2025). TinyML-enhanced cloud-edge collaborative framework for real-time sport action recognition. *Internet Technology Letters*, *8*(5), e70100.

Yan, Y., Chen, Q., Nasir, R., Bone, C., & Tuan, W.-J. (2025). Characterizing patients at higher cardiovascular risk for prescribed stimulants: Learning from health records data with predictive analytics and data mining techniques. *Computers in Biology and Medicine*, *188*, 109870.

Yang, B., Wang, L., & Sun, B. (2025). Advances in self-powered electrochemical systems for sports applications: Harnessing energy and sensing. *International Journal of Electrochemical Science*, *20*(8), 101080.

Zhao, Y., He, H., Tong, J., Li, W., & Zhou, S. (2025). Application of nanogenerators in lumbar motion monitoring: Fundamentals, current status, and perspectives. *Diagnostics*, *15*(20), 2657.

Zhu, P., & Hu, Y. (2025). Carbon nanomaterials intelligent wearable devices for real-time athlete monitoring and performance tracking. *Revista Materia*, *30*, e20250327.

Zim, M. K. I., & Hanif, M. A. (2025). Predictive analytics in mental health: AI for identifying risk factors and preventing crises. In *Proceedings of the 6th IEEE International Conference on Recent Advances in Information Technology (RAIT 2025).*

7 Computer Vision and Image Analytics for Industrial Safety

7.1 DEEP LEARNING FOR PPE ADHERENCE TRACKING

Personal protective equipment (PPE) compliance is part of occupational safety management in hazardous industries. Every worker wears necessary protective gear, such as helmets, gloves, masks, safety vests, and goggles. It traditionally depends on manual inspection or auditing of events after they have taken place. Limitations in human oversight because of fatigue, cognitive overload, and subjective judgement have driven rapid adoption. Computer vision-based systems enabled by deep learning (DL) algorithms, especially those based on convolutional neural networks and transformer architectures, can automatically analyse the visual content of workplace scenes captured by surveillance cameras to detect proper wearing of PPE items. It also quantifies compliance over time to support proactive safety management based on data evidence. Modern architectures like YOLOv8, EfficientDet, and Vision Transformers further raise the level of accuracy attained even in complex industrial scenarios characterized by variable lighting, occlusion, and overlapping motions.

The general logic for operational PPE adherence systems contains image acquisition, preprocessing, feature extraction, object detection, and compliance classification. Cameras installed across factory floors can continuously capture real-time video streams. They are fed into a deep learning model, which is capable of identifying classes for specific PPE. Large annotated datasets showing the appearance variance of the workers, colour, angle, and size of PPE are constructed during model training. Transfer learning with pre-trained weights on large-scale datasets such as COCO or ImageNet lends speed to convergence and overcomes the problem of limited domain-specific samples. The trained model makes inferences regarding the presence or absence of target PPE components in new frames with high accuracy. In addition, algorithms for temporal smoothing can be added, such as recurrent layers, in order to verify whether the adherence of PPE persists over several frames, considering this might lower the false alarms due to transient occlusion or motion blur. It is not just about substituting periodic inspections, but it also gives safety teams the capability to identify systematic patterns of noncompliance in certain departments or shifts by means of dynamic visual recognition (Mumtaz et al., 2025; Zhang et al., 2026a; Zhang et al., 2026b).

PPE tracking using deep learning should be adapted to the environment of an industrial setting for practical deployment. For example, the reflective gear worn by workers in metallurgical plants will confuse the visual recognition, while in the case of pharmaceutical production, masks and gloves have variations in material

DOI: 10.1201/9781003754213-9

transparency. Such variations require fine-tuning of the detection model through domain adaptation techniques. The creation of synthetic data with the aid of GANs to simulate camera angles, PPE colour, and partial occlusion has become an important complement to scanty datasets. Once more, multimodal fusion, integrating visual data with either RFID or thermal sensors, can enhance the robustness of the overall system. In the event that the ID badge of a worker, having a linked RFID chip, is identified in an unsafe zone without confirmation by the vision model about his or her wear of PPE, the system could flag a potential violation automatically. Such a fusion mechanism can increase robustness under conditions of low visibility or image noise (Hanafi et al., 2025; Lievens et al., 2025; Yang et al., 2025).

The best example of such a mature level of deep learning for PPE adherence has to be the case of Protex AI and Marks & Spencer. The Protex Vision System relies on CNNs, fine-tuned from thousands of factory-specific images, to detect safety noncompliance that could be a missing helmet, inappropriate footwear, or an unfastened safety vest. Then it goes beyond mere detection: it calculates a compliance score, summing up the observations of time, workers, and zones. The scores show up on dashboards accessed by managers and safety officers, thus allowing longitudinal monitoring and evidence-based decision-making. Specific patterns, like frequent omission of helmets in the loading dock, trigger focused retraining or policy intervention. Therefore, the feedback loop between AI analytics and organizational learning mechanisms leads to a measurable reduction in safety incidents. Protex AI clearly shows exactly how deep learning-based safety analytics can move from reactive supervision towards predictive governance.

Samsung Electronics provides a complementary example where PPE tracking forms part of a more comprehensive cyber-physical safety ecosystem: deep learning algorithms operating on continuous video streams at the Suwon smart manufacturing facility monitor adherence to PPE and safe interaction with robotic systems. At each workstation sits an AI-enabled camera hooked up with the in-house-developed safety management platform, which aggregates event data and performs predictive analytics. When deviations are noticed like the lack of gloves around an automated welding robot, the system immediately alerts supervisors. It is done via mobile notifications while logging the event for later statistical processing. Perhaps more important is how feedback from these analytics underpins ergonomic optimization, highlighting recurring posture risks or patterns of overexertion. Indeed, it is such holistic integration that underlines how PPE adherence systems can evolve into wider human-machine safety intelligence frameworks.

Therefore, methodologically, the correct detection of PPE depends on the interaction among such variables as illumination conditions, camera resolution, and heterogeneity in the PPE design. Additionally, high-intensity glare due to reflective material deceives segmentation layers, while dark-coloured or transparent protective gear may be misclassified as background. The implementation of adaptive thresholding and exposure correction algorithms at the preprocessing stage, together with model training based on data augmentation, allows for better generalization. Attention mechanisms integrated into CNNs or transformer blocks enhance the focus on the salient regions of PPE and hence provide their precise localization among complex industrial backgrounds. Empirical studies have shown that such a hybrid CNN-transformer system can achieve mAP greater than 95% in a controlled test environment; real-world deployment typically yields slightly lower metrics because of environmental noise (Ada et al., 2025; Kwangsukstith et al., 2025; Li et al., 2025b; Mao et al., 2025; Zhai et al., 2025).

While there is clear technological maturity, deep learning for PPE adherence introduces several organizational and ethical considerations. Continuously monitoring employees may raise concerns around privacy, especially in jurisdictions that have tight regulations in this regard, such as the GDPR in the European Union. Furthermore, ethical deployment requires anonymization of personal identifiers and strict governance in terms of data storage, access, and retention. Moreover, algorithmic bias needs to be overcome – if the model were trained based on data reflecting just a small subset of the PPE configurations, it would fail in detecting valid alternate forms of protection. Similarly, reporting mechanisms should be transparent, and human oversight in the loop is warranted. Today, many industrial organizations supplement AI-generated insights with manual review protocols in a way that will guarantee interpretability and accountability when enforcing safety (De Oliveira et al., 2025; Huang & Cheng, 2025; Majumder & Chatterjee, 2025; Talikowska et al., 2025).

Deep learning has transformed PPE adherence monitoring by introducing scalable, objective, and adaptive surveillance that translates visual information into actionable safety intelligence. These systems increasingly interface with industrial facility digital twins, thus enabling dynamic and risk-scenario simulations of compliance. Results from computer vision analytics become part of a continuous improvement ecosystem when integrated with Enterprise Resource Planning or environmental, health, and safety software. Things like real-time recognition, predictive analytics, and ethical AI governance – the hallmarks of convergence – are a shift from compliance auditing to intelligent prevention. It is an evolution central to the future of Industry 5.0. In it, human safety is not just observed but algorithmically safeguarded in harmony with technological autonomy (Al-Mhdawi & O'Connor, 2026; Lucena et al., 2026; Reis et al., 2025; Vaňková et al., 2025).

7.2 REAL-TIME VIDEO STREAM FOR UNSAFE BEHAVIOUR DETECTION

Real-time video analytics have changed the paradigm in managing occupational safety. The ability to automatically identify unsafe acts as they occur, especially in industrial settings where human error and behavioural negligence continue to top the list of common causes of accidents, is a major technological and managerial breakthrough. Deep learning-based computer vision systems currently work not just as passive monitoring but rather as intelligent sentinels that interpret motion, posture, and context to make out hazardous behaviour patterns such as running in restricted zones, incorrect manual handling of heavy objects, unauthorized use of equipment, or dangerous proximity to robotic arms. It detects deviations from normative behaviour models through continuous video stream analysis and automatically triggers alerts for timely intervention, hence reducing the time between risk emergence and response (Chang et al., 2023; Tian et al., 2025).

At the technical core of real-time unsafe behaviour detection are coupled temporal deep learning architectures and models that can analyse not only individual frames but also the dynamics of movement sequences. Together with more traditional CNNs, there are RNNs, LSTM units, or increasingly transformer-based architectures that can learn spatiotemporal dependencies. It is a model that is supposed to take in sequences of frames and extract both appearance and motion information on how human motion evolves in time. These models will be trained on large datasets of labelled behaviours – safe and unsafe – eventually learning normal operational conduct from hazardous deviations. The operator leaning too far over a conveyor belt may be classified as a potential falling hazard, while employees entering a zone without authorization may be detected via posture trajectory with contextual mapping (Febiyani et al., 2025; Feng & Cao, 2024; Li et al., 2025a).

This will involve a multistage analytics pipeline on video streams, whereby, firstly, object and person detection usually occurs through YOLOv8, Faster R-CNN, or EfficientDet-type models to detect people and any other relevant industrial entities like machinery, tools, and barriers inside every frame. Secondly, pose estimation is performed using algorithms such as OpenPose or MediaPipe, which estimate key skeletal points describing postures and orientations of limbs. The pose vectors resulting may be further processed through spatiotemporal networks, such as 3D CNNs or temporal graph convolutional networks, for the purpose of classification of behavioural patterns. As would be the case in the concrete example, this would also detect activities related to walking, bending, lifting, or slipping, and it would go ahead by inferring ergonomic correctness or possible fatigue indicators. These raw video streams are thus transformed into structured behavioural intelligence, with multiline fusions of detections, pose estimations, and temporal analyses that enable immediate safety alerts and allow long-term ergonomic analysis.

The above case of Boeing-EksoVest thus sets an implicit precedent for such dynamic monitoring. In even more developed forms, the detection of unsafe behaviour runs in line with adherence analysis of PPE as part of the same real-time safety management system, such as at Samsung Electronics' Suwon facility. AI-powered cameras screen the constant flow of assembly lines and robotic workstations to identify poor posture, gestures indicative of fatigue, or dangerous proximity relative to automated equipment through models. If it detects a deviation, further contextual validation is done. Because of that, it can make sure, for instance, that the motion is not part of an approved procedure. Also, a prioritized alert is sent to supervisors through an internal communication dashboard. This kind of real-time feedback loop reflects a shift from retroactive auditing to predictive and pre-emptive safety governance (Chen & Qian, 2023; Jia et al., 2025; Liu et al., 2025a).

Besides simple static camera feeds, multi-camera fusion and edge computing are being utilized in deploying industrial safety more and more. Two cameras positioned orthogonally around a workstation can identify whether or not a worker is inside the restricted area or just near it. These edge devices are small processors embedded near the camera and can perform compute-intensive inferences right at the place of image capture to reduce latency and bandwidth. Applications of this architecture include the issuance of alerts within milliseconds, something quite important in high-risk industrial contexts such as chemical processing, mining, or heavy robotics. Many such edge devices run optimized versions of neural networks, quantized or pruned for efficiency, with aggregated results sent to central servers for analytics, visualization, and policy integration.

Classifying unsafe behaviours in real-time video systems is far from trivial, requiring not only the accuracy of visual recognition but also contextual understanding of the operational workflow. For example, bending can be safe while one is retrieving a tool but unsafe while being performed under an active conveyor. Recent advances use context-aware learning, training models with semantic scene segmentation and task metadata to associate behaviours with specific zones, tools, and operation conditions. GNNs are especially good at this. It can be used in modelling spatial relationships between the worker, objects, and machine parts in a scene. These are combined with rule-based safety logic. Also codified from occupational safety standards such as ISO 45001. Such systems can infer whether a detected action constitutes a violation of safety protocols or an acceptable variation within task parameters (Park et al., 2023; Zhang et al., 2023).

Of the existing challenges in real-time unsafe behaviour detection, false positives are some of the most persistent: rapid movements, temporary occlusions, or tool reflections may be misclassified as violations. Hybrid architectures have been put to work to reduce such errors by embedding deterministic rule checking into probabilistic inferences from the outputs of deep learning. The general contribution of ensemble methods is high reliability because they pool over several models. Temporal smoothing and thresholding on the confidence will also filter out the noise since no alarm will fire without sustained evidence of violation. At times, human feedback loops are implemented where supervisors have access to manual labelling of false alerts, and through continuous learning, the model will regularly be retrained to refine its precision over time. This is the symbiosis of AI autonomy and human oversight. It ensures operational accuracy and user trust (Ji et al., 2024; Ying et al., 2024).

Visualization is a core part of such behavioural analytics. Advanced industrial dashboards (those from Protex AI or IBM Visual Insights) can visualize heat maps, timelines of violations, and trajectories of workers. These allow the safety manager to identify the emergence of spatial or temporal clusters of unsafe behaviour, such as repeated unsafe lifting at a given workstation or increasing fatigue indicators during night shifts. The data-driven approach thus moves safety management away from intuition-based supervision towards quantitative behavioural intelligence. Third, the aggregated video analytics inform organizational learning; repeated deviations may point to a lack of training, poor ergonomics of the workstations, or systemic scheduling stressors. In this way, the system does not just detect individual violations but uncovers deeper structural inefficiencies in the production ecosystem.

Real-time video analytics integrated into the broader safety infrastructure requires a multitier data architecture wherein edge devices perform localized inference, factory servers aggregate event data and apply rule-based reasoning and statistical filtering, and enterprise cloud-based platforms store historical records and produce predictive insights. Trained on accumulated behavioural data, predictive models are able to forecast those periods or areas that will experience especially high risk, for example, shifts with high overtime rates or production zones plagued by frequent equipment failures. This vertical integration of perception (video stream), cognition (deep learning inference), and management (safety decision support) constructs a closed-loop control system for industrial safety, well aligned with the cyber-physical systems logic of Industry 5.0 (Önal & Dandıl, 2025; Soumith et al., 2024).

The ethical dimension of real-time behavioural monitoring cannot be sidestepped. Continuous surveillance naturally invokes questions of privacy, consent, and even its psychological impact on workers. While the life-saving intent of the system is beyond doubt, such systems can easily create an atmosphere of hyper-observation. The line between safety and dignity is tenuous and requires careful design in terms of governance mechanisms. Other best practices constituting ethical deployment involve anonymization of video feeds, access to data restricted to authorized personnel only, and deletion of raw footage once analytic extraction has taken place. Secondly, it is important that workers be clearly informed of the extent and purpose of monitoring, with participatory mechanisms through which they can express their concerns or suggestions. Ethical frameworks inspired by core principles of responsible AI – fairness, accountability, transparency, and explainability – should therefore permeate both the design and operation of such systems.

Empirical evidence from early industrial adopters has underlined the effectiveness of real-time unsafe behaviour detection. Incident rates at those facilities where AI-driven monitoring has been deployed show measurable declines, often by 20–40% within the first operational quarter. This does not come merely from immediate responses to alerts but also from a process of behavioural adaptation: workers, aware of intelligent oversight, internalize safer practices. Yet, the long-term sustainability of such behavioural change depends on how the organization can complement surveillance with positive reinforcement and safety culture development. AI monitoring should, therefore, be positioned as supportive intelligence and not some punitive authority. With periodic reviews, training reinforcement, and collaborative feedback mechanisms between human supervisors and AI analytics, real-time detection can serve as a foundation for continuous improvement and organizational learning (Li et al., 2025a; Wu et al., 2025).

The summary of the key considerations – technological, operational, and ethical – relating to real-time video analytics for unsafe behaviour detection within an industrial environment is captured in the summary shown in Table 7.1. It is about the state-of-the-art deep learning architecture, epitomized by 3D convolutional and transformer networks, combined with pose estimation and temporal modelling, semantically grasping complex human activities across dynamic workspaces, to multi-camera fusion and edge computing that enables on-spot responses at low latency, a must for high-risk profile operations. Then there is context-sensitive detection, which consolidates system intelligence by linking detected behaviours with environmental and procedural data. Adaptive threshold and ensemble methods further suppress false positives, increasing reliability. Predictive analytics enable proactive risk management. It emerges from continuous monitoring. That surfaces recurring unsafe patterns or fatigue-related behaviours. Ethical governance is pursued through compliance with GDPR, anonymization of data, and consultative communication with employees. It is warranted that such systems' deployment will improve the safety culture without any violation of human dignity and autonomy (Anjana et al., 2025; Bonyani et al., 2024; Wang et al., 2025; Wang & Mou, 2025).

TABLE 7.1
Principal Technological, Operational, and Ethical Components of Real-Time Video Analytics for Unsafe Behaviour Detection

Aspect	Technical Implementation	Industrial Application/ Example
Motion and posture Recognition	It does so by analysing sequential frames to identify human body dynamics, using 3D CNNs, LSTM, or Vision Transformers.	Detection of improper bending or twisting during assemblies in automotive plants.
Pose estimation	Keypoint extraction: OpenPose or MediaPipe for skeletal mapping of workers.	Monitoring of lifting posture in logistics centres to prevent musculoskeletal injuries.
Object and person detection	Deployment of YOLOv8 or Faster R-CNN for real-time identification of workers, machinery, and hazardous zones.	Automatic detection of human presence around the robotic welding arms.
Temporal behaviour analysis	Integration of temporal modelling to differentiate between transient and sustained risky actions.	Recognizing prolonged unsafe leaning over conveyors or repetitive strain gestures.
Multi-camera fusion	Synchronized views from several cameras are combined via 3D spatial modelling to eliminate blind spots.	Multi-angle safety monitoring in large-scale semiconductor manufacturing.
Edge computing integration	Perform ultra-low latency response data processing locally to avoid delays in cloud transmission.	Immediate triggering of alarms at chemical plants upon handling dangerous materials.
Context-aware detection	Linking behaviour to environmental semantics: zones, equipment type, and task phase.	Differentiate normal maintenance entry from unauthorized access to restricted areas.
False positive mitigation	Adaptive confidence thresholds, ensemble inference, and human-in-the-loop verification.	Reducing false alarm activations in poor lighting conditions for metallurgical sites.
Predictive safety analytics	Aggregating behavioural data over time can serve to predict risk hotspots and fatigue patterns.	Identifying high-risk shifts with frequent unsafe motions at electronic assembly lines.
Ethical and governance framework	Application of GDPR, anonymization, and transparent communication with employees.	Ethical deployment of monitoring systems in European smart factories under ISO 45001 and AI Act.

Source: Author's own work.

7.3 ANNOTATION, FALSE POSITIVES, AND ETHICAL DEPLOYMENT

The efficiency and trustworthiness of industrial safety computer vision lean on the three interconnected dimensions: quality of data annotation, FPs and FNs control, and ethical principles related to deploying a system. Such dimensions form the cognitive, operational, and normative backbone of artificial safety intelligence. Deep learning models differ from classic rule-based detectors because the ability to recognize something does not come from explicit logic but rather from data representation. Therefore, their accuracy and fairness relate to how images and behaviours are labelled during training, to methodologies used for validation of the detection accuracy, and to frameworks governing how such technology will interact with human autonomy and privacy. This subchapter will develop these dimensions in order, moving from data-centric aspects of annotation to algorithmic calibration for FP reduction to socio-ethical considerations defining sustainable and legitimate deployment in an industrial context (Skubis et al., 2024).

Data annotation is an inherently interpretive act in which raw visual material gets transformed into structured knowledge for machine learning. In applications to industrial safety, this generally involves the manual or semi-automatic labelling of workers, equipment, and protective gear, postures, and behaviours across thousands of video frames. For example, even for training a simple model to detect whether a worker is wearing a helmet, each and every frame must have bounding boxes around the region of interest of the head with a categorical label like ‘helmet’ or ‘no helmet’. The annotation required to detect unsafe behaviour can get even more complicated, since it requires temporally coherent annotation – that is, identifying not just objects but motion sequences signifying risk, such as bending while lifting, leaning over an edge, or stepping into a restricted zone.

High-quality annotation is inherently multidisciplinary in occupational contexts, requiring coordination between safety engineers, data scientists, and annotators. The interpretive ambiguity of human action builds in a risk of bias: what one expert considers unsafe behaviour might be contextually appropriate for another. For this reason, annotation protocols should be standardized, with clear ontologies for safety-relevant events. These will typically flow from occupational standards such as ISO 45001, OSHA guidelines, or local regulations concerning safety. A number of annotation tools like Computer Vision Annotation Tool (CVAT), LabelMe, or Supervisely allow for consistent labelling of bounding boxes or segmentation, while version-controlled datasets make every labelling decision traceable and auditable.

Recent breakthroughs in active learning and semi-supervised annotation have eased the high labour intensity of the process. Active learning iteratively selects uncertain samples for human review, hence focusing the annotation effort where it really matters. Semi-supervised frameworks leverage pseudo-labelling, where the model’s own high-confidence predictions are used as provisional annotations to extend the dataset. Synthetic data generation, especially with GANs, is increasingly used to create those rare scenarios – slipping or colliding workers – which are either hard or unsafe to record in real environments. Nevertheless, even synthetic data requires validation by humans to ensure semantic realism. The quality of annotation directly impacts model generalization. It contains a lack of diversity, which leads to overfitting. Low consistency in labelling results in ambiguous decision boundaries and a higher false detection rate (Dalila & Turkben, 2025; Marciniak et al., 2025; Rosak-Szyrocka & Wolniak, 2025; Srinivasan et al., 2025; Wolniak & Stecuła, 2024).

In safety-critical AI, the FPs and FNs have asymmetrical consequences. While an FP-classification of a safe behaviour as unsafe might lead to operational disruptions, spurious warnings, and user fatigue, an FN-failure to recognize a true hazard can result in injury or death. The nature of model calibration means there is always a trade-off between sensitivity and specificity with respect to the risk context. In a high-stakes setting, such as one with hazardous chemicals, a heightened tolerance towards FPs would be viable to minimize the potential for FNs. Conversely, scenarios where continual false alarms lead to operational disruption demand much tighter thresholds (Karaduman & Yildirim, 2025; Lloyd-Roberts et al., 2025).

Threshold optimization, ensemble learning, and context-aware inference are some of the technical methods to reduce FPs and FNs. It adjusts the confidence level at which a model issues a detection. Adaptive thresholds will change dynamically depending on zones or the time of day. It is to reflect contextual risk intensity. Ensemble learning averages out the biases of various models by combining predictions from several architectures, such as CNN, transformer, and LSTM models. Besides, temporal smoothing requires many frames with non-overlapping behaviour to pass in order for them to be classified; hence, transient errors due to occlusion or motion blur are mitigated. Further removal of physically impossible detections should be possible through geometric constraints-based post-processing filters, such as plausible joint angles based on pose estimation (Al Rawajbeh et al., 2025; Vijay Anand et al., 2025).

Most operational FP management strategies rely on feedback loops between human supervisors and the AI: where there has been an alert that was confirmed or dismissed by human review, the system archives the outcome and periodically refines its internal decision surfaces. Continuous learning turns the AI from a static classifier into an adaptive observer of real-world patterns. Predictive analytics on FP distribution may spot systematic weaknesses – frequent misdetections in low light or near metallic surfaces – and drive targeted data augmentation. Transparency in error reporting is key: explainability modules, which visualize which regions of the image influenced a given decision, now form part of many industrial AI dashboards, helping safety officers assess whether a detection was rational or spurious (Liu et al., 2025b; Ma et al., 2025).

One has to adopt not only a strictly technical focus on error rates but also take into consideration the psychological dimension of trust; for if a system produces too many false alarms, then ‘alert fatigue’ may set in, which is a euphemism for the fact that human operators either quit paying attention or disable the system altogether. The calibration process, therefore, needs to be informed by human factors engineering, as well as by precision-recall metrics. The interface should clearly convey confidence levels, differentiate minor from critical alerts, and provide contextual information about the detected event. Ethical responsibility goes beyond algorithmic performance to make sure automation enhances, rather than burdens, human situational awareness (Guo et al., 2025).

Industrial deployment of computer vision systems touches fundamental principles of autonomy, privacy, fairness, and accountability. Unlike traditional safety sensors, a vision-based AI system observes not only the physical condition but also human behaviour and identity, making working spaces continuous areas of surveillance. A number of important questions are hence raised regarding ownership of visual data, the length of its storage, its possible reuse for other purposes, and whether opting out is possible for workers. In jurisdictions falling under the GDPR, such systems represent a form of biometric data processor with a lawful basis for processing data, data minimization, and explicit consent or justification based on legitimate interest.

The need to balance such safety objectives with ethical obligations through the taking of a privacy-by-design approach is considered by many organizations. For example, video feed processing can be performed at the edge, and the raw footage need never leave the local device; only abstracted metadata, such as ‘helmet missing, zone 3’, is transmitted to central servers. Furthermore, face and body anonymization through blurring or silhouette rendering can prevent the identification of individuals. Strict access controls, encryption, and automatic data-deletion policies should also be put in place in systems. However, these technical safeguards do not suffice, and transparent governance structures should be set up if the deployment of these systems is ethically legitimate. Workers’ councils or safety committees should be involved in defining the scope of monitoring and incident analytics reviews and in assessing whether the degree of surveillance is proportional to gains in safety (Cano et al., 2026; Chintalapati et al., 2025).

Other key aspects of ethical deployment involve algorithmic fairness. Models predominantly trained on specific demographic and physical attributes may show bias. For instance, detection of PPE adherence is notably less accurate across different body shapes, skin tones, or cultural variations in protective attire. Results in penalization or false accusations can thus be disproportionate and, therefore, damaging to fairness and trust. This goes in line with the emerging EU AI Act and ISO/IEC 42001 on AI management systems, where countermeasures will involve collecting balanced data, auditing bias, testing explainability, and recording how data are collected, annotated, and used, including accountability mechanisms in case of harm, maintained by all involved organizations. Ethics impact assessments should also be retained by all organizations involved through documentation of how data are collected, annotated, and used, setting accountability mechanisms in case of harm (Brandt et al., 2025; Naderalvojoud et al., 2025).

Ethical deployment also involves consideration of psychological and organizational repercussions: continuous monitoring changes the dynamics of the workplace, creating anxiety or even distrust. In such a case, implementation must be coupled with open communication, training, and participatory dialogue to avoid such a situation. Workers have to see AI not as an instrument of discipline but as an ally in maintaining health and well-being. This will mean a cultural shift in which safety is understood to be a matter of collective responsibility supported by smart tools, rather than imposed from above by an impersonal algorithm. The concept ‘augmented supervision’ captures this synergy: humans retain decision authority, while AI provides evidence-based insights to enhance situational awareness (Suriyakumar et al., 2025; Teng et al., 2025; Vogt, 2025).

Maturity implies convergence on three dimensions: high-quality annotation, calibration of FP/FN, and ethical governance. Each dimension forms a nexus with others: high-quality annotation reduces model bias and false detections; robust calibration ensures operational usability, while ethics support social legitimacy. The next wave is probably going to be federated learning, a model trained collaboratively across different sites, with no actual exchange of raw data, while keeping confidentiality, but with better generalization. Given the advancements in explainable artificial intelligence (XAI), explanation frameworks will be further developed that offer intuitive visualizations explaining why the system labelled an action as unsafe.

Ethical deployment means translation of technical compliance into moral intelligence, a quality of systems that could align technological capability with human values. At its core, Industry 5.0 knits together human dignity, autonomy, and ingenuity along with sophisticated automation. Computer vision is more than a surveillance tool; it is a reflective medium through which industrial cultures negotiate between efficiency and safety, then between safety and freedom. If AI-driven safety analytics are well-governed, they elevate the culture of the workplace by replacing punitive monitoring with cooperative prevention, turning visual data into the language of trust, care, and shared responsibility (Kolla et al., 2026; Suk & Han, 2025; Williams Tolbert, 2025).

Table 7.2 summarizes the key ethical and legal dimensions framing responsible deployment in applying AI-driven video analytics to industrial safety management, from compliance with privacy and data protection legislation through transparency

TABLE 7.2
Ethical and Legal Dimensions that Frame the Responsible Deployment of AI-Driven Video Analytics in Industrial Safety Management

Dimension/Principle	Description and Industrial Implication
Privacy and data protection	Personal visual information must be captured and treated in accordance with the principles enshrined in GDPR and ISO 42001 on minimum retention and anonymization.
Lawful basis for processing	Organizations also have the responsibility to justify the usage of information by either legitimate interest or explicit consent, well-recorded legal grounds for monitoring.
Anonymization and de-identification	They include face blurring, rendering in silhouettes, or masking bodies to avoid the identification of workers as individuals.
Data minimization	Only metadata from safety events, like 'helmet missing', should be stored. Raw footage is deleted after analytic extraction.
Transparency and communication	Accessible policies on the purpose, extent, and use of the system should be made available to workers, including regular briefings.
Algorithmic fairness and bias control	Training datasets also have to include a wide array of body types, skin tones, and differences in personal protective equipment to prevent discriminatory outcomes.
Explainability and accountability	Systems should integrate modules of explainable AI that will enable decision logic visualization and audit trails for each detection.
Human oversight and decision authority	Human agency and contextual reasoning are preserved through a human supervisor making the final call on AI alerts.
Worker participation and governance	Safety committees or workers' councils should be co-designers of the monitoring policies, and thus they should assess proportionality between surveillance and safety gain.
Security and access control	In this case, video data has to be encrypted with graded access privileges and secure protocols for transmission against unauthorized use.
Ethics impact assessment	Organizations should consider periodic evaluation of continuous monitoring's psychological, cultural, and organizational effects.
Proportionality and social legitimacy	It should be proportionate to the real occupational risk, finding a reasonable balance between safety and dignity.

Source: Author's own work.

and participatory governance, in addition to GDPR and ISO 42001. Each of these dimensions, from anonymization and data minimization through fairness, explainability, and human oversight, has been considered in the framework such that technological innovation remains second to ethical integrity and social legitimacy. Mechanisms for algorithmic accountability, associated with secure data handling and bias control, protect trust in organizations and the rights of individuals. In addition, proportionality holds: the intensity of surveillance must correspond to the level of actual danger and not to excessive control, which would erode autonomy or morale. Together, they form a normative framework in which computer vision systems are not instruments of supervision but are ethically framed as part of a human-centred culture of safety.

BIBLIOGRAPHY

Ada, Y. R., Surono, A., Supriyati, S., & Pribadi, A. P. (2025). Strategies to improve compliance with personal protective equipment use in the textile industry: A scoping review. *BMC Public Health*, *25*(1), 2743.

Al Rawajbeh, M., Maria Soosai, A. J., Ramasamy, L. K., & Khan, F. (2025). Trustworthy adaptive AI for real-time intrusion detection in industrial IoT security. *IoT*, *6*(3), 53.

Al-Mhdawi, M. K. S., & O'Connor, A. (2026). Impact of sustainable safety climate on poor compliance with personal protective equipment in oil and gas construction projects: Quantitative analysis and best sustainable practices. *Environmental Impact Assessment Review*, *117*, 108211.

Anjana, P., Shibi, B., Vinoth Kanna, I., & Ajayakumari, K. (2025). AI-powered real-time accident detection through integration of driver health and vehicle dynamics. In *Proceedings of the International Conference on Trends in Engineering Systems and Technologies (ICTEST 2025)*.

Bonyani, M., Soleymani, M., & Wang, C. (2024). Construction workers' unsafe behavior detection through adaptive spatiotemporal sampling and optimized attention based video monitoring. *Automation in Construction*, *165*, 105508.

Brandt, L., Au, L., Castro, C., & Odom, G. J. (2025). Engaging an advisory board in discussions about the ethical relevance of algorithmic bias and fairness. *NPJ Digital Medicine*, *8*(1), 292.

Cano, F., Henzinger, T. A., & Kueffner, K. (2026). Algorithmic fairness: A runtime perspective. *Lecture Notes in Computer Science*, *16087*, 1–21.

Chang, J., Zhang, G., Chen, W., Yuan, D., & Wang, Y. (2023). Gas station unsafe behavior detection based on YOLO-V3 algorithm. *China Safety Science Journal*, *33*(2), 31–37.

Chen, X., & Qian, K. (2023). A case of computer image processing depth learning algorithm based on YOLOv4 improved algorithm. In *Proceedings of the IEEE International Conference on Image Processing and Computer Applications (ICIPCA 2023)* (pp. 1134–1138).

Chintalapati, A., Enkhbat, K., Annamalai, R., Ozaydin, F., & Noel, M. M. (2025). Quantum-enhanced algorithmic fairness and the advancement of AI integrity and responsibility. *Quantum Reports*, *7*(3), 36.

Dalila, R. A. M., & Turkben, A. K. (2025). Artificial intelligence based partial discharge detection using CNN and KNN to increase the quality of electrical insulation. *Discover Computing*, *28*(1), 101.

De Oliveira, F., Sant'Ana, H. G., De Freitas, F. H., Chandler Velemem Alves, J., & Humberto Miranda, E. (2025). Isometric muscle endurance, linear acceleration, and change-of-direction ability: What are the effects of the personal protective equipment used by special military forces? *Journal of Bodywork and Movement Therapies*, *44*, 745–749.

Febiyani, A., Suhardi, B., Laksono, P. W., & Prastawa, H. (2025). Predictive model approach to enhancing occupational health based on safety culture and sustainable technology in environmental, social, and governance. *Media Publikasi Promosi Kesehatan Indonesia*, *8*(9), 933–947.

Feng, Z., & Cao, T. (2024). FOA-BDNet: A behavior detection algorithm for elevator maintenance personnel based on first-order deep network architecture. *AIMS Mathematics*, *9*(11), 31295–31316.

Guo, J., Zhang, R., & Wang, J. (2025). Optimized YOLOv5 framework for real-time detection of aircraft fuel sediments in aviation maintenance systems. *Advances in Transdisciplinary Engineering*, *74*, 802–813.

Hanafi, A., Refdinal, & Irfan, D. (2025). Personal protective equipment (PPE)-detector: A lightweight automatic detection of protective gears in construction sites based on YOLO architecture. *ICIC Express Letters*, *19*(10), 1063–1070.

Huang, M.-L., & Cheng, Y. (2025). Dataset of personal protective equipment. *Data in Brief*, *62*, 111988.

Ji, Z., Zhou, Y., Zhang, Y., Guo, X., & Shi, K. (2024). Industrial site unsafe behavior detection based on improved YOLOv5. *China Safety Science Journal*, *34*(7), 38–43.

Jia, X., Zhou, X., Shi, Z., Xu, Q., & Zhang, G. (2025). GeoIoU-SEA-YOLO: An advanced model for detecting unsafe behaviors on construction sites. *Sensors*, *25*(4), 1238.

Karaduman, G., & Yildirim, H. (2025). Descriptor-based explainable QSAR modeling approaches for the prediction of respiratory toxicity of volatile and gas-phase chemicals. *Journal of Applied Toxicology*, *45*(11), 2437–2449.

Kolla, A. C., Takale, D. G., Mahalle, P. N., Sule, B., & Deshmukh, G. (2026). Mitigating algorithmic bias in facial recognition technology: A deep learning approach to fairness in AI. *In Lecture Notes in Networks and Systems*, *1457*, 423–438.

Kwangsukstith, S., Surawattanasakul, V., Mahakkanukrauh, C., Panumasvivat, J., & Sirikul, W. (2025). Assessment of occupational hazards, health effects, and personal protective equipment usage among motorcycle food delivery riders in Thailand: A cross-sectional survey. *Journal of Occupational Medicine and Toxicology*, *20*(1), 13.

Li, H., Wu, L., Zhong, X., Guo, L., & Cui, Y. (2025a). Recognition of personnel fatigue state and unsafe behavior based on computer vision. *China Safety Science Journal*, *35*(3), 28–35.

Li, Z., Wang, Y., Shang, Y., & Lu, Y. (2025b). Urban pandemic governance personal protective equipment allocation strategies: A system dynamics simulation. *Scientific Reports*, *15*(1), 3452.

Lievens, S., Ricci, L., Dumont, G., Velimirovic, M., & Merlino, S. (2025). From COVID-19 personal protective equipment (PPE) to pollutants: A multi-analytical approach towards environmental weathering of gloves and face masks in beach sand and sea. *Marine Pollution Bulletin*, *221*, 118530.

Liu, Y., Wang, P., & Li, H. (2025a). An improved YOLOv5s-based algorithm for unsafe behavior detection of construction workers in construction scenarios. *Applied Sciences (Switzerland)*, *15*(4), 1853.

Liu, Z., Pan, X., Zhang, S., & Cai, F. (2025b). Industrial anomaly detection method by fusion of self-coding feature extraction and continuous immunization mechanism. In *Proceedings of the 6th International Conference on Computer Information and Big Data Applications (CIBDA 2025)* (pp. 925–931).

Lloyd-Roberts, B., Pantekis, F., James, P., O'Reilly, L., & Edwards, M. (2025). Scaling invariant generation using state space embeddings and GPU streaming. *Concurrency and Computation: Practice and Experience*, *37*(25–26), e70353.

Lucena, R., Almeida, N., & Simões, P. (2026). Physiological impact of personal protective equipment in distinct thermal environments. *Lecture Notes in Computer Science*, *15888*, 105–121.

Ma, C., Gong, J., & Gan, X. (2025). FPGA-accelerated deep learning for millisecond-level safety enforcement in tobacco manufacturing. *Advances in Transdisciplinary Engineering*, *74*, 987–996.

Majumder, A., & Chatterjee, S. (2025). YoloGA: An evolutionary computation-based YOLO algorithm to detect personal protective equipment. *Journal of Intelligent and Fuzzy Systems*, *49*(5), 1251–1264.

Mao, Y., Zhu, Y., Feng, X., & Fang, Z. (2025). Development of heat stress index for healthcare workers with personal protective equipment. *Energy and Buildings*, *346*, 116150.

Marciniak, K., Majewski, P., & Reiner, J. (2025). Inspection of railway catenary systems using machine learning with domain knowledge integration. *Scientific Reports*, *15*(1), 29426.

Mumtaz, H., Sobek, S., Sajdak, M., Muzyka, R., & Werle, S. (2025). Optimizing advanced oxidative liquefaction of municipal solid waste and personal protective equipment of medical sector for solid reduction and secondary compounds production. *Renewable Energy*, *255*, 123831.

Naderalvojoud, B., Curtin, C., Asch, S. M., Humphreys, K., & Hernandez-Boussard, T. (2025). Evaluating the impact of data biases on algorithmic fairness and clinical utility of machine learning models for prolonged opioid use prediction. *JAMIA Open*, *8*(5), ooaf115.

Önal, O., & Dandıl, E. (2025). Unsafe-Net: YOLO v4 and ConvLSTM based computer vision system for real-time detection of unsafe behaviours in workplace. *Multimedia Tools and Applications*, *84*(29), 34967–34993.

Park, M., Tran, D. Q., Bak, J., Kulinan, A. S., & Park, S. (2023). Real-time monitoring of unsafe behaviors of portable multi-position ladder workers using deep learning based on vision data. *Journal of Safety Research*, *87*, 465–480.

Reis, A., Silva, V., Joaquim, J. J., Mateos-Campos, R., & Moreira, F. (2025). Health effects of ergonomics and personal protective equipment on chemotherapy professionals. *Current Oncology*, *32*(10), 563.

Rosak-Szyrocka, J., & Wolniak, R. (2025). *AI-Driven Sustainability: The Future of Human Resources Management* (p. 224). Taylor & Francis Group.

Skubis, I., Wolniak, R., & Grebski, W. W. (2024). AI and human-centric approach in smart cities management: Case studies from Silesian and Lesser Poland Voivodships. *Sustainability*, *16*, 18.

Soumith, B. S., Lochan, K., Murthy, S. N., Vishwanath, M. N., & Venugopal, N. (2024). *Driver monitoring system*. In *Proceedings of the 15th International Conference on Advances in Computing, Control and Telecommunication Technologies (ACT 2024)* (Vol. 2, pp. 3267–3272).

Srinivasan, S., Ramesh, T. K., & Paccapeli, R. (2025). Safety integrity level (SIL) assessment process for WSN using multi-QoS metrics. *MethodsX*, *15*, 103660.

Suk, Y., & Han, K. T. (2025). Evaluating intersectional fairness in algorithmic decision making using intersectional differential algorithmic functioning. *Journal of Educational and Behavioral Statistics*, *50*(5), 833–862.

Suriyakumar, V. M., Zink, A., Hightower, M., Ghassemi, M., & Beaulieu-Jones, B. (2025). Computational challenges arising in algorithmic fairness and health equity with generative AI. *Nature Computational Science*, *5*(9), 698–700.

Talikowska, M., Belcher, J., Golling, E., Birnie, T., & Finn, J. (2025). The quality of CPR delivered by EMS personnel wearing enhanced personal protective equipment during the COVID-19 pandemic: A retrospective cohort study from Perth, Australia. *Resuscitation Plus*, *26*, 101062.

Teng, Z., Xia, H., & He, Y. (2025). Algorithmic fairness and digital financial stress: Evidence from AI-driven e-commerce platforms in OECD economies. *Journal of Theoretical and Applied Electronic Commerce Research*, *20*(3), 213.

Tian, F., Song, C., & Liu, X. (2025). Research on miner's unsafe behaviour detection algorithm based on improved AlphaPose. In *Proceedings of the 5th International Conference on Artificial Intelligence and Industrial Technology Applications (AIITA 2025)* (pp. 589–593).

Vaňková, E., Obrová, K., Hodek, J., Weber, J., & Scholtz, V. (2025). Scaled-up non-thermal plasma-generating device Plasmatico enables effective and harmless disinfection of personal protective equipment. *Scientific Reports*, *15*(1), 35631.

Vijay Anand, R., Magesh, G., Alagiri, I., Kesavan, M., & Abdullah, A. B. (2025). Industrial Internet of Things cyber threats detection through deep feature learning and stacked sparse autoencoder based classification. *Transactions on Emerging Telecommunications Technologies*, *36*(9), e70224.

Vogt, Y. (2025). Disability and algorithmic fairness in healthcare: A narrative review. *Journal of Medical Artificial Intelligence*, *8*, 56.

Wang, F., Wang, H., Fu, R., & Zhang, X. (2025). Study on unsafe behavior detection of tower crane drivers in prefabricated building construction. *IEEE Transactions on Intelligent Transportation Systems*, *26*(4), 4406–4417.

Wang, H., & Mou, L. (2025). An improved YOLOv8 based unsafe behavior detection algorithm for coal mine underground personnel. In *Proceedings of the 6th International Conference on Computer Engineering and Application (ICCEA 2025)* (pp. 180–184).

Williams Tolbert, A. (2025). Why causal inference is necessary for algorithmic fairness. *Synthese*, *206*(3), 162.

Wolniak, R., & Stecuła, K. (2024). Artificial intelligence in smart cities: Applications, barriers, and future directions—A review. *Smart Cities*, *7*, 3.

Wu, Y., Feng, S., Wu, Y., Hu, Z., & Wu, C. (2025). Masked region disparity-based unsafe behavior detection via campus monitoring device images. *Knowledge-Based Systems*, *310*, 113051.

Yang, W., Kim, M. S., Sippel, G. J., Marsic, I., & Burd, R. S. (2025). An image dataset for surveillance of personal protective equipment adherence in healthcare. *Scientific Data*, *12*(1), 96.

Ying, L., Lei, Z., Junwei, G., Lei, M., & Zilong, Z. (2024). Unsafe behaviour detection with the improved YOLOv5 model. *IET Cyber-Physical Systems: Theory and Applications*, *9*(1), 87–98.

Zhai, N., Tang, R., Huang, Z., Qin, J., & Huang, Y. (2025). Assessment of hospital environmental and personal protective equipment contamination with influenza A and B viruses by ultrasensitive droplet digital PCR. *Diagnostic Microbiology and Infectious Disease*, *113*(3), 116963.

Zhang, J., Liu, C.-C., & Ying, J. J.-C. (2023). DeepSafety: A deep neural network-based edge computing framework for detecting unsafe behaviors of construction workers. *Journal of Ambient Intelligence and Humanized Computing*, *14*(12), 15997–16009.

Zhang, Q., Xu, R., Zhang, Z., Xu, X., & Li, J. (2026a). Medical staff personal protective equipment detection network via global information and multi-path feature fusion. *Communications in Computer and Information Science*, *2622*, 374–386.

Zhang, Q., Yang, L., Qi, Y., Li, Q., & Miao, R. (2026b). Skeleton-based detection of anomalous personal protective equipment doffing behaviors among healthcare workers. *Journal of Safety Science and Resilience*, *7*(1), 100229.

Zheng, X., Wang, T., Chowdhury, S. Y., Sun, R., & Luo, D. (2023). Unsafe behavior detection with adaptive contrastive learning in industrial control systems. In *Proceedings of the 8th IEEE European Symposium on Security and Privacy Workshops (EuroS&PW 2023)* (pp. 363–369).

8 Real-Time Localization and Obstacle Avoidance

8.1 RTLS FOR TRACKING EMPLOYEE AND MACHINE MOVEMENTS

Real-time locating systems have rapidly become the foundational element in modern industrial settings, especially in facilities characterized by dense human-machine interactions. This is representative of the greater move towards continuous spatiotemporal intelligence, in which static safety assumptions are replaced by dynamic monitoring of flows, proximity relations, and emergent risks. The industrial workplace is no longer perceived as a fixed physical setting; rather, it is a complex cyber-physical space in which workers, robots, autonomous vehicles, and mobile equipment move around continuously, constantly reshaping spatial configurations. RTLS technologies create the informational substrate underpinning this novel form of understanding. Thanks to the accurate localization, high-frequency data acquisition, and AI interpretation of movement patterns, organizations can coordinate resources, avoid collisions, make ergonomic assessments, and optimize task allocation. This subsection gives an overview of RTLS architecture, sensing modalities, operational capability, quality implications, and integration challenges (Burggräf et al., 2024; Jonek-Kowalska & Wolniak, 2025; Patrick Sullivan et al., 2023; Rosak-Szyrocka & Wolniak, 2025b; Wolniak & Stecuła, 2024).

RTLS conceptually shifts workplace supervision from an observation/human-controlled monitoring paradigm to that of distributed sensor-driven computation tracking. Safety in manufacturing operations has traditionally rested on combinations of prescriptive norms, mechanical barriers, and visual vigilance. Such strategies are bound to fall short when mobile robots travel down shared corridors, forklifts need to traverse narrow aisles, and employees themselves frequently change location to perform their tasks. RTLS introduces a consistently updated map of the environment, enabling both central systems and edge devices to draw inferences about movements and distances. It therefore provides, in effect, a spatial information layer atop the physical facility.

What sets RTLS apart from other data collection is the real-time attribute. While status information may come mainly from inventories, processes from warehouse management systems, production traceability modules, or ERP systems, RTLS offers one flow of dynamic positional updates. The transition from static information to fluid spatiotemporal data sets will have two immediate consequences: first, it will enable organizations to detect anomalies at the very moment they occur, a very critical aspect in terms of safety and continuity of operations; second, it will allow them to create a detailed retrospective trajectory, hence enabling investigations, optimization studies, and risk analyses (Hayes et al., 2024; Li et al., 2024; Yang et al., 2023).

DOI: 10.1201/9781003754213-10

RTLS is based on several sensing technologies whose choice depends on environmental conditions, required accuracy, possibility of interference, and cost structures. Ultra-wideband solutions allow very accurate ranging by measuring the time-of-flight of radio pulses. In collision-avoidance and proximity warning systems, positional accuracy in the range of 10–30 cm might suffice. BLE beacons are appropriate for large-scale facilities when cost efficiency is needed, although their precision is lower. Both passive and active RFID-based systems are required in those applications where identification needs to be added to tracking. Vision-based localization is increasingly more and more accurate thanks to recent results in computer vision and deep learning, but it introduces issues related to occlusion, variability of lighting conditions, and constraints linked to privacy. Hybrid architectures show increasing relevance: ultra-wideband combined with inertial measurement units or BLE with machine vision may provide higher robustness in complex environments (Benouakta et al., 2023; Wang et al., 2023; Zhang et al., 2025).

It normally comprises location sensors or anchors installed in fixed locations, mobile tags attached to workers or equipment, and a central localization engine. Advanced deployments can also include edge computing modules for pre-processing the signals locally for reduced latency. In highly demanding environments, the refresh rate can be as high as 10–20 Hz, allowing the system to respond quickly to fast changes in motion and relative distances. Besides, positioning accuracy is further improved by machine learning and correcting for multipath effects or anchor geometry calibration based on historical data.

The most important industrial driver for the implementation of RTLS is risk reduction within mixed traffic zones, that is, space shared between humans and machines. In such settings, forklifts, AGVs, and mobile robots have heterogeneous movement dynamics: their acceleration behaviours are different, their turning radii vary, and so do their braking characteristics. Classical safety protocols assume given pathways or fixed separation zones – neither of which holds in those flexible production systems centred around lean flows and on-demand material delivery. RTLS allows for dynamic spatial negotiation whereby the system continuously evaluates if a trajectory of a machine intersects with the forecasted trajectory of a worker (Bhattacharya, 2025; Cordero, 2025; Kim & Park, 2024; Sidiropoulos et al., 2025).

In such a context, raw position signals are processed by AI algorithms to get the movement vectors, velocity estimates, and predictions of future states. In this respect, the system will continuously observe not only the current distance but also closing speed, relative orientation, and the probability of path convergence. Because of this, it will be possible to raise a local alarm, for example, to the operator through haptic or acoustic signals, or to the machine itself, which can then perform autonomous emergency braking in case of threshold exceedance. The literature increasingly underlines the effectiveness of such systems in reducing near-miss incidents and lowering the cognitive load on workers who no longer need to surveil their surroundings continuously (Cao et al., 2023; Danh et al., 2023).

RTLS allows the performance of ergonomic assessments. Worker trajectory reconstruction makes it possible to determine factors such as exposure to zones of high risk, burdens of repetitive movements, and inadequate task distribution. Supplemented with wearable sensors measuring motion or load, RTLS offers a multidimensional view of human activity targeted for intervention.

Some of the key topics of RTLS implementation concern integrations with MES, WMS, and digital twins. In modern architectures, RTLS is positioned as one of the main data producers that feed enterprise AI. Real-time localization becomes part of the contextual information needed for predictive scheduling, automatic task dispatching, and smart material handling. For instance, a production line digital twin can take RTLS data to simulate the impact of personnel mobility on cycle times or test different layouts using this data.

RTLS often represents a bridge between OT and IT in most facilities, capturing events in the physical space and relaying them to the analytical engines driving insight. It needs to be able to ensure secure communication, assurance of synchronized time stamps across systems, and adherence to data governance frameworks. Latency management will be critical: delayed localization data undermines collision-avoidance logic. This will require industrial-grade networks, including Wi-Fi 6 or private 5G, to achieve the bandwidth and reliability required (Krugh et al., 2024; Sayfoori et al., 2025; Volpi et al., 2023a).

RTLS adoption thus brings quantifiable changes in both industrial efficiency and safety. The risk of collisions decreases, while near-miss incidents may be reduced as much as eightfold. Organizations can better visualize mobile resources and, therefore, help in resource allocation, reduction of time spent idly, and avoidance of bottlenecks. RTLS can communicate which equipment is unused, or it can optimize forklift routes based on actual movement patterns rather than based on assumed workflows (Kim et al., 2024; Ochoa-de-Eribe-Landaberea et al., 2024).

Spatial behaviour transparency acts on organizational culture. Workers will feel much safer in knowing that intelligent systems support hazard detection, while sensitive treatment is warranted around privacy considerations. Continuous tracking of staff calls into question the valid reasons for monitoring. Ethical deployment requires clear communication of the practice and limitation of data retention at the individual level with explicit separation between safety analytics and performance surveillance being also important. Indeed, industrial standards are increasingly underpinning anonymization and the use of movement metrics as aggregated values rather than individual identifiers.

RTLS simplifies training programmes. Research into the regular pathways of an experienced operator allows an organization to identify the best practices and embed them within the training modules. Novice employees will benefit from contextualized instructions regarding how experts get through a complex setting.

Next-generation RTLS solutions now integrate machine learning and real-time analytics that will shift from reactive alerts towards predictive spatial intelligence. While traditional RTLS provides the location, AI-driven RTLS interprets behaviour – algorithms that classify motion modes like walking, running, reversing, and turning pick up anomalous trajectories indicative of a higher risk threshold. Reinforcement learning models optimize safe routing in dynamic environments. Situational maps integrated with AI systems update risk zones according to fluctuating traffic density (Berkhan et al., 2024; Pilati et al., 2024; Sherif & Ahmed, 2024).

Table 8.1 summarizes operational benefits and their organizational implications, eloquently depicting RTLS's transformation effects within an industrial setting by entwining real-time spatial intelligence with measurable performance outcomes. Each benefit, from reduced near-miss events to automated proximity alerts, translates into

TABLE 8.1
Operational Benefits and Organizational Implications of RTLS Deployment

Operational Benefit	Organizational Implication
Reduction of near-miss events	Strengthening of safety culture and a measurable decline in incident rates
Precise tracking of forklifts and mobile robots	Improved coordination among human operators and autonomous systems
Real-time visibility of worker movements	Improved spatial planning and redesign of traffic routes
Optimization of equipment utilization	Reduction of idle times and better allocation of mobile assets
Dynamic risk-zone adjustment	Transition from static to adaptive safety management
Data-driven analysis of congestion points	Smarter layout optimization and flow engineering
Early detection of hazardous trajectories	Faster decision-making in safety-critical situations
Automated proximity alerts	Reduction of cognitive load on workers and machine operators
Reconstruction of movement histories	More accurate root-cause analysis of operational disruptions
Integration with MES/WMS systems	Greater coordination of logistics and production activities
Enhanced monitoring of ergonomic exposure	Targeted interventions for workload balancing and fatigue prevention
Support for training and competency development	Standardization of best practices based on expert movement patterns

Source: Author's own work.

wider organizational impacts on safety culture, stability in workflows, management of ergonomic risk, and strategic resource allocation. This table underlines the fact that RTLS is not just a location-tracking technology but one in which an organization reshapes how it understands mobility patterns, coordinates human-machine interaction, and drives data-informed enhancements at both levels of the production and logistics system. The table shows the systemic nature of the adoption of RTLS in Industry 4.0 settings, from immediate operational gains to longer term organizational shifts.

Predictive models anticipate congestion points. For example, if multiple forklifts head towards a material retrieval area at roughly the same time, the system will sequence their access to avoid delay and unsafe build-ups of traffic. Predictive path planning has particular value in fleets of autonomous robots where navigation decisions consider human presence and ever-changing environmental constraints. AI helps contribute to sensor fusion by analysing inconsistencies between UWB signals, inertial measures, and environmental mapping. The system can remove noisy data and reconstruct more accurate trajectories. This is highly important in environments with metal structures, reflective surfaces, or electromagnetic interference (Burggräf et al., 2024; Ghafoorpoor Yazdi et al., 2024; Valiollahi et al., 2024; Tebaldi et al., 2025).

Thus, from the organizational point of view, the cost is a limiting factor mainly for small and medium-sized enterprises (SMEs). BLE or hybrid solutions reduce this price, though very highly accurate UWB systems remain expensive. Maintenance, replacing batteries within mobile tags, and periodic recalibrations add to operational costs.

Among all the factors, privacy is the most important aspect. The legal restrictions on personal data mean that the storage, access, and processing are strictly limited. The organizations involved in this sector very often adopt either pseudonymization or real-time processing to avoid long-term archiving of individual trajectories in whatever form.

Success with RTLS requires user acceptance. Workers must perceive the system as a safety tool and not as a way to monitor them. The alerts must be appropriately designed; too many warnings result in habituation and eventual ineffectiveness. The human-machine interface needs to balance sensitivity against noise suppression (Berkhan et al., 2024; Shamsollahi et al., 2024; von Wagner et al., 2024).

Table 8.2 identifies a series of challenges and limitations in balancing constraints around the implementation of RTLS with wider organizational implications. It is in this way that technical problems of interference, environmental variability, or integration difficulties arising interact with human, financial, and regulatory factors in the adoption landscape along several dimensions. The investment in network

TABLE 8.2
Challenges and Limitations of RTLS and their Organizational Consequences

Challenge or Limitation	Organizational Consequence
High initial installation cost	Budget constraints and slower adoption in the SME environment
Complex calibration of anchors and sensors	Specialized technical abilities and long implementation periods are required
Signal interference in metal-dense facilities	Reduced the accuracy of localization and hybrid sensing requirements
Occlusion caused by machinery and infrastructure	Reduced the reliability of tracking workers in the critical zone
Battery dependence of wearable tags	Higher maintenance workload and periodic shutdowns for replacements
Data latency in large-scale facilities	Lower responsiveness of the collision-avoidance logic
Integration challenges with legacy MES/WMS systems	Development of additional middleware and interoperability issues
Privacy and data protection concerns	The need for strict governance, anonymization, and building trust among employees
Over-alerting and false positives	Fatigue from warning signals and reduced system credibility
Resistance to adoption by workers	Need for communication strategies and training programmes
Requirement for robust wireless infrastructure	More investment in upgrading the network to Wi-Fi 6 and private 5G
Environmental variability (temperature, dust, vibration)	Reduced sensor life and more frequent maintenance cycles

Source: Author's own work.

infrastructure, the creation of strong data governance frameworks, and proactive communication among employees are some of the structural responses that should be made to a number of individual limitations and may be helpful in mitigating concerns related to monitoring and privacy. In making that explicit distinction, successful RTLS deployment depends upon technological accuracy, organizational readiness, capability for long-term maintenance, and ethical and compliance expectations being met.

Trio Mobil, for instance, carries out advanced deployment of RTLS for Mercedes-Benz. The manufacturer faced challenges related to complex flows of workers and forklifts in high-density areas of assembly. Fixed safety zones could not reflect variable patterns of operation, while during high-activity periods, the risk of collisions goes up.

Trio Mobil's AI-based RTLS revolutionizes safety management by providing high-precision localization for both forklifts and worker tags, feeding a predictive risk-mapping engine. It continuously estimated movement vectors, thus bringing to light potential points of intersection and allowing early warnings. Most impressively, the formerly problematic intersections became collision-free. Dynamic resizing of the risk zones was done with respect to traffic density, evidently an example of how spatial intelligence can adaptively fit real operational contexts rather than rigid, predefined boundaries.

Mercedes-Benz benefited not only from the improvements in safety but also from the operational optimization: smoothened forklift flows, reduced unnecessary idle times, and the system allowed data-driven redesigns of aisle structures. Generally, this case proves that RTLS, combined with AI, is much more than tracking, acting like an operational intelligence platform to support the continuous improvement and efficiency of resources.

RTLS is part of the infrastructural layer of Quality 4.0. High-resolution location data amplifies the accuracy of safety management while shrinking variability in logistical flows, driving evidence-based decision-making. It is the spatial data in quality-oriented organizations that nurture root-cause analyses of disruptions, ergonomic assessments, and predictive maintenance routines for mobile equipment. While digital twins have become a very common element in industrial operations, it is RTLS that provides the real-world grounding that those virtual models need to stay in sync with physical dynamics.

Spatial tracking extends this philosophy of process transparency, whereby quality management becomes more dependent on immediate observation of material and human flows to locate inefficiencies or latent hazards. RTLS amplifies this transparency to include microdynamics in day-to-day operations.

RTLS technologies surely represent one step toward intelligent, adaptive, and safety-oriented industrial environments. Integrated with AI, the possibility of applying predictive reasoning to spatial dynamics turns workplaces into a coordinated ecosystem of humans and machines. Many technical, financial, and ethical challenges notwithstanding, their strategic value is great. It is continuous spatial intelligence that modern production and logistics systems call for, and RTLS provides the mechanism for generating it.

8.2 MODELS OF COLLISION PREDICTION AND MAPPING OF RISK ZONES

This has driven the development of sophisticated models of collision prediction and dynamic risk mapping, spurred by growing densities of mobile entities within industrial spaces: workers on foot, fleets of autonomous mobile robots (AMRs), forklifts, and articulated robotic platforms. These form a computational layer interpreting spatial data in real time and inferring the likelihood of hazardous interactions before they actually happen. While RTLS offers positional updates, collision-prediction frameworks turn raw spatial information into structured assessments of risk. This transformation is crucial for safety management in high-mobility factories where even a single delay in hazard recognition can lead to a serious accident. Rather than merely reactive alerts, the main purpose of these models is to derive anticipatory intelligence – they infer trajectories, classify movement intentions, quantify spatial uncertainty, and update themselves with every change in conditions inside the production environment.

Collision-prediction models have traditionally combined kinematic analysis with probabilistic reasoning and machine-learning methods. Traditional approaches use velocity, orientation, and acceleration to rigidly predict trajectories under the assumption of linear continuations of motion. As such, they are inherently suited for rigid machines with well-defined and predictable behaviour, whereas they can be quite insufficient in capturing spontaneity and variability in humans. Modern approaches model uncertainties using stochastic processes. Bayesian filters, particle filters, and Kalman-based methods have been widely explored in attempts to estimate future positions either in the presence of noisy or missing data or in situations characterized by irregular motion patterns. Probabilistically, a system is able to analyse not only an expected trajectory but also a range of plausible deviations (Deva Hema & Rajeeth Jaison, 2024; Han et al., 2024; Park et al., 2025; Yuan et al., 2025).

Besides the level of kinematics, sophisticated models today are increasingly including behavioural prediction. Human locomotion can be understood using methods ranging from crowd dynamics and social-force models to the learning of motion primitives. These methods all reflect the insight that movement is a function of spatial intentions, obstacle avoidance strategies, and task-specific patterns. For machines, the prediction involves constraints in the form of turning radii, braking curves, and collision-avoidance protocols embedded into AMR controllers. As industrial spaces move to collaborative autonomy, models will have to balance heterogeneous motion types within a coherent risk assessment framework.

Trajectory forecasting is the backbone of collision-prediction systems. Real-time forecasting engines process sequences of positional data and infer short-term futures. The most common forecasting horizon ranges between 1 and 5 seconds; this is a window long enough to prevent collisions but short enough to remain computationally manageable. These models are being increasingly applied to this task using machine-learning-based models, mainly recurrent neural networks such as LSTM and GRU, and temporal convolutional networks. They learn patterns of acceleration, deceleration, and directional changes that cannot be captured by static equations (Atif et al., 2025; Jurkus et al., 2025).

Forecasting performance is contingent upon many factors. Data density dictates how well one can infer velocity and acceleration; sparser updates lead to higher uncertainty. Environmental structure also informs predictions: narrow aisles, blind corners, or gate zones narrow down the possibilities for movement that models must account for explicitly. The difficulty further scales with the presence of multiple interacting agents. In these settings, forecasting becomes inherently a multi-agent problem that involves jointly modelling the interdependent trajectories. Graph neural networks and attention mechanisms are some of the emerging tools to represent such relational dependencies.

Once the trajectories are predicted, the next step is the computation of collision probability. Computation here identifies whether two trajectories will cross each other within a spatiotemporal window. Analytical approaches compute the minimum distance between forecasted paths. If the minimum forecasted distance, after accounting for object size, speed, and uncertainty, is lower than the threshold, the system flags the situation as potentially unsafe. Probabilistic approaches extend this reasoning by incorporating confidence intervals around position forecasts. Then, the intersection of the probability distributions among agents signals collision likelihood.

In real-life implementations, thresholds cannot be static; they must respond to context. For example, a low threshold would suffice in the low-speed pedestrian area, while crossings at high speeds need more conservative thresholds. Therefore, adaptive thresholding models take into consideration such environmental conditions, time of day, task intensity, and traffic density. Such contextualization ensures that alerts are neither too sensitive nor dangerously delayed (Hafid et al., 2025; Makhdoomi et al., 2025; Zhong et al., 2025).

The spatial constructs of risk zones define areas of increased hazard. They are developed by overlaying environmental structure with temporal patterns of movement. Static risk zones originate from facility design and include, but are not limited to, the following: intersections, blind spots, bottlenecks, loading docks, and machine operation areas. Dynamic risk zones are based on real-time activity. High forklift traffic, concentrations of workers, or sudden changes in the flow of traffic produce temporary expansions of risk areas.

These new mapping engines use spatial-temporal data fusion, merging 2D and 3D facility maps with position streams to refresh heatmaps many times a second. These resulting maps choreograph safe routing, instruct the behaviour of autonomous vehicles, and deliver contextually relevant alerts. Of prime significance, though, risk-zone mapping helps plan, not just avoid, hazards. By observing how the risk zones shift through different production cycles, it is possible for an organization to reshape layout structures, adjust the scheduling, or reposition machines to reduce exposure to hazards in the longer term.

Artificial intelligence underpins risk mapping by highlighting emergent patterns that remain invisible to rule-based systems. Clustering algorithms highlight areas where deviances from normal movement occur, such as unexpected congregations of workers or peculiar equipment behaviour. Deep-learning models analyse spatio-temporal sequences to make inferences on unseen trends correlating with increased risks, including fatigue patterns, repetitive task loops, or changed movement behaviours within peak hours (Han et al., 2024; Rosak-Szyrocka & Wolniak, 2025a; Skotnicka-Zasadzień & Wolniak, 2026; Skubis et al., 2024; Tao et al., 2025; Wolniak & Stecuła, 2024).

Models can be of great value in dynamic environments. Using AI systems analysing historical data on movement, estimates may be made when an area is going to become congested. They estimate bottlenecks that may occur from planned production orders, expected material-delivery times, or typical worker schedules. The predictive capability thus allows pre-emptive re-routing and prevention of hazardous traffic density formations.

Semantic mapping techniques continually refine the models of risk zones. Identifying the function of industrial spaces, such as assembly, logistics transfer, inspection, and maintenance, automatically creates systems that generate risk classification maps corresponding to both movement density and operational context. Machines operating in the maintenance bays would need different risk buffers than those navigating logistics corridors.

Key applications of collision-prediction models include integrated utilization with AMRs and AGVs, as these machines need real-time situational awareness to go through the motions in industrial areas safely and effectively. Collision-prediction algorithms are built into the control architecture of these machines; thus, they can offer adaptive path adjustments, immediate braking, or even cooperative routing with human workers.

Shared-control scenarios are, therefore, an area of rapid growth. In shared-control scenarios, humans and machines together contribute to the avoidance of collisions: machines adjust their behaviour based on the prediction of human trajectories, and workers get warnings or visual cues that steer them onto safe trajectories. This bidirectional awareness greatly reduces the possibilities of high-energy collision events.

Another aspect of integration concerns human–robot collaboration cells. At such collaborative workstations, the robot and a human work directly in proximity with each other. Collision models track limb movements and predict reaching behaviours to adjust robot speed or operating volume. This dynamic modulation replaces fixed safety fences and extends flexibility to hybrid workflows.

The most relevant issue related to collision prediction is how to handle uncertainty: Localization signals can degrade because of multipath interference, occlusions, or signal reflections. The machines could have unexpected mechanical delays, and the humans may move along in non-standard ways for numerous cognitive or task-related reasons. Because of this, collision-prediction systems directly embed uncertainty in their computational logic.

The primary mechanism for reducing uncertainty is sensor fusion. UWB, BLE, infrared, lidar, machine vision, and inertial measurement units produce robust spatial estimates through the process of data fusion. Most of the fusion architectures are based on weighted confidence scores, Bayesian optimization, or deep-learning-based fusion layers. This type of method provides resistance against any single sensor failure and continuity of the predictions when one modality is temporarily degraded.

Uncertainty propagation models further quantify the influence that localization noise has on collision probability by accounting for the confidence intervals of the predicted position and, correspondingly, tuning safety buffers. In such a manner, the approach ensures that alerts remain reliable under nonideal conditions (Bhattacharya, 2025; Cordero, 2025; Zhang et al., 2025).

The effective deployment of any collision-prediction model requires strict validation, typically including performance metrics on the accuracy of the prediction – that is, how closely forecasted trajectories actually match real movement. It also includes false-alarm rates because system reliability directly influences user trust; reaction times or speed at hazard identification; spatial precision, which quantifies actual risk zones against real-world hazard locations; and lastly, operational impact metrics that analyse the effects of collision prediction on productivity, machine idle time, and worker flow stability.

Also, operational validation shall be performed under conditions of maximum traffic, maintenance, emergency drills, and shift changes. Performance shall be stress-tested for robustness with noise, occlusion, and high-density interactions.

The applications of collision prediction and risk-zone mapping in industrial settings will span the domains of logistics warehouses to intermodal terminals. Real-time traffic forecasting reduces congestion and allows for safer pallet transfer operations in logistics warehouses. Predictive risk mapping will make possible the dynamic speed modulation of AGVs that deliver components along automotive assembly lines. Collision models in ports and intermodal terminals protect workers navigating through stacks of containers and coordinate autonomous equipment operating along transfer routes.

It helps manage maintenance zones in process industries where technicians and heavy machinery come together through predictive mapping. The classification of risk zones also integrates safety-critical constraints against hazardous materials in chemical and pharmaceutical works. All these applications confirm that collision prediction is indeed not solely a safety mechanism but a core component of operational orchestration.

They directly support the strategic objectives of Quality 4.0 and safety management by making it possible to proactively mitigate risks, reduce variability in operational flows, and feed into continuous-improvement cycles. Data from collision-prediction systems inform quality analytics, root-cause investigations, and layout optimization processes. They enhance digital twins with dynamic behavioural patterns and improve simulation environments with higher fidelity.

More broadly, predictive risk mapping enhances organizational resilience: the ability to anticipate rather than respond to hazards reduces downtime for an enterprise while improving reliability and assuring continuity of workflow. These effects influence both operational stability and long-term capability development.

Collision-prediction models represent a significant development in industrial safety and mobility management; they provide an interpretation of complex industrial dynamics by merging real-time data, probabilistic reasoning, machine learning, and environmental semantics. Their role goes beyond hazard detection into design, human-machine coordination, and strategic decisions regarding layouts. Collision-prediction models will represent one of the fundamental enabling mechanisms in the adaptive, safe, and efficient industrial ecosystems of highly autonomous and interconnected factories in the future (Atif et al., 2025; Jurkus et al., 2025; Park et al., 2025; Yuan et al., 2025).

8.3 DYNAMIC EMERGENCY ALERTS AND SITUATION AWARENESS OF THE ENVIRONMENT

Industrial settings are rapidly becoming contexts of high mobility, multisource data flows, and interaction between human activity and autonomous machine operations. Classic mechanisms of emergency response – manual alarms, fixed sirens, and static evacuation plans – no longer suffice in such contexts based on predictable hazards and linear escalation patterns, considering that today's factories and logistic centres operate on fluctuating workloads, changing layouts, and dynamic movement patterns. Dynamic emergency alert systems will strive to bridge this gap by interpreting real-time data streams, assessing contextual risk factors, and issuing adaptive warnings reflecting the transient state of the industrial ecosystem. What is beneath this discourse is an understanding that emergencies are not considered a binary issue but rather dynamic processes that need continuous interpretation and concerted reaction.

Dynamic architectures of emergency alerts integrate sensing technologies, decision-support algorithms, and multimodal communication channels. This is rooted in an understanding that static alarms can capture no temporal complexity in industrial hazards. A chemical leak will spread unevenly according to conditions of airflow, configurations of equipment, or shifts in personnel. A fire in a densely automated warehouse would spread differently as robots relocate stock or the ventilation systems react to the heat. Alert systems must integrate environmental variables, motion signals, and equipment diagnostics. In this way, emergency notification moves beyond a simple binary trigger to become an evolving situational model (Feng et al., 2025; Zhang et al., 2025).

At their centre lies the notion of situational awareness, perhaps defined as the state of continuous perception, comprehension, and projection of environmental states. Computationally, situational-awareness calls for real-time fusion of data streams emanating from thermal cameras, gas detectors, RTLS devices, vibration sensors, machine-embedded diagnostics, and digital-twin simulations. This fused dataset would be used by the alert logic to classify the severity, identify the local vulnerabilities, and recommend actions to workers, supervisors, and autonomous machines. Today, emergency alerts are tailored, targeted, and time-sensitive.

What distinguishes a dynamic alarm system is that it treats hazards in context, not as some kind of homogeneous event. Detection of a temperature increase, for example, should in itself not constitute sufficient cause for a full facility alarm. Its spatial position, probable ignition sources, and availability of combustible material in its immediate surroundings, weighed against manpower concentration in this area, are considered. The detection of gas undergoes a similar type of reasoning: concentration gradients, flow patterns of air, and structural containment affect the degree of severity classification. This way, warnings can be contextualized to stay relevant and therefore reduce the likelihood of alarm fatigue.

Dynamic assessment has several challenges, including the temporal evolution of hazards: the system has to detect anomalies but also needs to infer their probable trajectory – that is, predictive models will be estimating whether fire propagation is accelerating or if structural vibrations are indicative of progressive mechanical failure. Dynamic assessment reduces unnecessary evacuations while enabling immediate action when true risks escalate (Dafhalla et al., 2025; Mohsin et al., 2025; Shadpey et al., 2025).

Instead of simple on/off alarms, there is differentiation between early warning, targeted local alerting, and full emergency notification through multi-level adaptive notification mechanisms. Early warnings supply situational cues to workers without interrupting tasks through displays on their wearables, subtle auditory tones, or status updates on machine displays. Depending on the intensification of the hazard or approaching sensitive zones, the system raises this communication by turning on localized alarms for workers in the immediate vicinity. When systemic danger is identified through predictive models, the system turns on facility-wide alarms.

The safety integrity with minimum operational disruption depends on this multi-layered structure. Warnings should be given through duplicate channels within an alert system for effectiveness in various environmental conditions. Wearable devices, voice annunciators, mobile apps, LED panels, and displays integrated into machines play complementary roles. The redundancy compensates for any failure in communication because of noise, power fluctuation, or network disruption.

Adaptive alerts support heterogeneous groups of users. However, production zone workers need only short and actionable messages, while supervisors and emergency coordinators must be supported by analytical dashboards that outline the evolution of hazards, affected zones, and recommended intervention strategies. Autonomous robots will obviously need machine-readable signals as a base to make them suspend operations, reroute paths, or transition into a safe state. These aspects are part of designing alerts targeted at such diverse audiences (Latha et al., 2025; Odilova et al., 2025).

The thermal camera's dynamic alert capabilities detect abnormal heat signatures, gas detectors track hazardous emissions, vibration sensors measure structural anomalies or failure precursors, and sound analysis systems capture abnormal acoustic patterns, such as sudden mechanical impacts or pressure releases. These sensors feed data in real time into predictive analytics engines to classify events and estimate escalation risks, possibly extending their impact downstream.

Indeed, the machine-learning models amplify such processes by finding complex patterns that conventional threshold-based systems cannot understand. For instance, deep-learning networks analyse thermal distributions to detect electrical faults at an early stage, long before a temperature would exceed a pre-set limit. Similarly, the reinforcement learning models select an optimal timing for the evacuation by taking into consideration simulated scenarios of human mobility and response times of equipment, as well as traffic congestion in the escape routes.

Another important advantage of predictive analytics is that it offers differentiation between false positives and actual emergencies. Noisy machinery, dust interference, and temperature fluctuations give very ambiguous sensor signals. Contextual machine learning (ML) models suppress irrelevant anomalies and highlight meaningful events, increasing system reliability.

These dynamic alerts create significant operational value from their integration with spatial awareness. Indeed, that is a result of combining RTLS and indoor mapping to make responses to emergencies an organized, spatial process. When a hazard is present, the system identifies all workers who are in the immediate and adjacent zones of risk; next, it provides customized guidance on safe egress routes, considering in real time the congestion of corridors, the position, and movement of mobile robots, besides that of the hazard itself.

The most sophisticated ability of dynamic systems is the optimization of escape routes. Traditional evacuation schemes rely on static diagrams that get printed and mounted on facility walls. In contrast, dynamic systems calculate personalized routes for each worker. Whenever corridors get blocked, a machine stalls in a narrow passage, or hazards spread unpredictably, these automatically change. Wearables or mobile devices will then give workers turn-by-turn instructions. Similarly, autonomous machines receive spatial instructions to ensure that any unplanned machine movements will not halt the flow of the evacuations (Devi et al., 2025; Kowshik et al., 2025; Monni et al., 2025).

Furthermore, it extends spatial intelligence to emergency services such that first responders entering the facility have continuously updated maps showing the development of hazards, distribution of workers, and blocked areas on which decision-making should be based with less uncertainty in critical moments.

The point is, however, that with the presence of AMRs, AGVs, and collaborative robots, the dynamics of emergency response change fundamentally: machines have to go into safe states, posing no additional risks under emergency conditions. Dynamic alert systems using industrial communication protocols signal machines to decelerate, park at defined positions, or free escape routes. Robots performing dangerous activities such as welding, cutting, and handling chemicals are obliged to shut off when receiving the signals at hazard levels.

One of the big challenges is that human-machine behaviours can be coordinated only insofar as machines are compatible with human responses, and humans must be advised about machine status. Currently, systems rely on shared situational-awareness models that ensure coherence for all agents. For instance, in commanding a robotic forklift to stop at an intersection in the corridor, human workers working in proximity are informed in parallel that the motion of the machines has been temporarily stopped so they would not form wrong assumptions about evacuation.

Moreover, robots can enable emergency interventions in hybrid workspaces. In such a scenario, the AMRs with payload platforms could transport specific equipment, deliver fire extinguishers, or assist persons with limited mobility. These examples of emerging applications underline the larger potential of dynamic alert frameworks to act as enablers of collaborative responses during emergencies.

More recently, digital-twin technologies started to play influential roles in dynamic alert systems. The reason is that, in this respect, the digital twin maintains a virtual model of the facility that is always up-to-date, in real time, and thus capable of performing simulation-based emergency evaluation. In the event of any kind of anomaly being detected, it starts building simulations on how that can evolve, considering airflow patterns, structural characteristics, and material distributions, including continuing machine operations. Correspondingly, these simulations provide early insights that will help in making the alert escalations but also inform strategies for emergency coordinators (Dani et al., 2025; Nivishna et al., 2025; Panda et al., 2025; Sneha et al., 2025; Vigram et al., 2025).

Digital twins also allow organizations to test emergency response procedures without disturbing production. Virtual drills simulate the movement of workers according to a host of hazard scenarios and measure evacuation times, highlighting various spatial or procedural weaknesses. Gained insights feed back into system calibration to refine alert logic and improve evacuation route algorithms.

Data from past events also enhances the digital-twin models. In this way, the integration of historical sensor data together with real-time input may turn the twin into a predictive engine that learns from operational history.

For all the technological sophistication, real-world effectiveness can only be as good as organizational preparedness. Workers need to understand how the alerts work, how to interpret multi-level warnings, and how to perform responsive action in time-sensitive situations. Training programmes must therefore involve practical simulations that are representative of realistic emergency scenarios.

The basis of designing alerts is very human factor oriented. For instance, complicated messages during emergency situations may mislead and fail to help. Alerts should be concise, direct, and unambiguous. Cultural influences are relevant in influencing user response, too. In some environments, the hierarchical pattern of communication may delay decision-making at the right time, while appropriate alert protocols should be designed to ensure action is taken on time, irrespective of organizational hierarchy.

Another crucial factor is trust: workers need to feel confident that the warnings will be valid in due time. Too many false alarms undermine credibility, while too few might also breed complacency. Thus, a delicate balance between sensitivity and specificity is crucial for system acceptance (Magar et al., 2025; Tundis et al., 2025).

Dynamic alert systems operate in a regulatory environment that prescribes occupational safety standards, data protection laws, and industry-specific compliance requirements. Sensor data used in the forecast of emergencies typically contain location information, patterns of movement of personnel, and even perhaps biometric data. It is up to the organization to ensure that data-minimization principles, retention limits, and access controls are upheld. Ethical considerations do not stop at the legal: workers must be fully informed about the processing of their personal data, and alert systems under no circumstances may turn into stealth employee-monitoring devices.

The modern landscape of regulations also increasingly appears to support adaptive safety mechanisms. For example, industrial safety now focuses on risk-based thinking; hence, dynamic alert systems can easily meet international norms such as ISO 45001. However, adherence will be ensured through documentation, validation protocols, and periodic auditing.

Such dynamic emergency alerts contribute to industrial resilience based on the speed with which they can act upon an unpredictable event in a data-driven manner. This will consequently provide strategic value with respect to continuity and protection of life, as well as minimizing equipment damage. In some facilities, even very short interruptions involve huge losses; here, dynamic alert systems work as stabilizing mechanisms that can stop cascading failures.

Beyond immediate safety, these systems allow the organization to learn from near misses and micro-incident events. Data collected from unusual events provides insight into structural vulnerabilities, behaviour patterns, and bottlenecks within emergency procedures. This instils a culture of continuous improvement over time and contributes to building high-reliability industrial settings (Dafhalla et al., 2025; Dani et al., 2025; Nivishna et al., 2025; Shadpey et al., 2025).

Table 8.3 summarizes how dynamic emergency alert and situational-awareness systems strengthen industrial resilience along several strategic dimensions. This is not, then, any kind of one-function solution but rather the result of mechanisms designed to work in concert for the protection of workers, the stabilizing of operations, and the provision for informed governance. On the one hand, dynamic alerts

TABLE 8.3
Strategic Role in Resilient Industrial Ecosystems

Strategic Dimension	Mechanism Enabled by Dynamic Alerts and Situational Awareness	Organizational Impact
Operational continuity	Early detection of disruptions and automated hazard escalation	Downtime is reduced, and production flows show more stability
Workforce protection	Personalized guidance and adaptive evacuation routing	Lower injury rates and a stronger safety culture
Asset preservation	Automatic shutdown of machinery in hazardous zones	Reduced damage to high-value equipment and infrastructure
Adaptive governance	Real-time visibility of evolving risks across departments	Faster decision-making and better coordination in emergencies
Learning from near misses	Collection of micro-incident data and anomaly patterns	Stronger continuous-improvement cycles and better root-cause analysis
Layout optimization	Identifying spatial bottlenecks and unsafe circulation paths	Informed redesign of facility structures and movement corridors
Robustness of autonomous systems	Synchronization between machine states and dynamic hazards	Higher reliability of AMRs, AGVs, and robotic platforms
Regulatory compliance	Automated documentation of alerts, responses, and environmental conditions	Easier audits, alignment with safety standards like ISO 45001
Organizational resilience	Maintenance of operational capability under abnormal conditions	Increased resilience to hazards, interruptions, and environmental stressors
Strategic decision support	Integration of alert data with digital twins and risk-simulation engines	Improved long-term planning, scenario analysis, and prioritization of investments

Source: Author's own work.

may enable disruption detection, safety responses at an individual level, and machinery shutdowns. Situational-awareness frameworks give real-time views on environmental states and the evolution of hazards. These have lessened downtime and injury rates, optimally preserved assets, and resulted in better adherence to regulations. For the organization, these reinforce continuous cycles of improvement, enable the redesign of facilities, and enhance the robustness of autonomous systems. Finally, integration with digital twins and simulation models through the use of data from alerts as input allows for long-range planning and strategic decision-making, all of which positions the modern safety infrastructures as foundational elements of resilient industrial ecosystems.

Dynamic emergency alerting and environmentally aware mechanisms represent a paradigm shift in how organizations understand safety. Such systems combine real-time data, predictive analytics, and spatial intelligence with human-centred approaches to building dynamic safety infrastructures adapted to evolving hazards. Everything but traditional alarm mechanisms, they will shape facility design, emergency planning, machine coordination, and organizational resilience. As industrial systems everywhere in the world become ever more connected, autonomous, and full of data, dynamic alerts will shape their safety and reliability.

BIBLIOGRAPHY

Atif, Y., Tarakanov, Y., Lebram, M., Karlsson, A., & Hemeren, P. (2025). Data-driven prediction of vehicle–vulnerable road user collisions at road intersections using machine learning models. *Procedia Computer Science*, *257*, 777–784.

Benouakta, A., Ferrero, F., Lizzi, L., & Staraj, R. (2023). Antenna characteristics contributions to the improvement of UWB real-time locating systems' reading range and multipath mitigation. *IEEE Access*, *11*, 71449–71458.

Berkhan, P., Kärcher, S., & Bauernhansl, T. (2024). Framework for the classification of real-time locating system (RTLS) use cases in matrix production systems. *Procedia CIRP*, *130*, 1346–1351.

Bhattacharya, S. (2025). Response to use of a real-time locating system in infection control. *Infection and Chemotherapy*, *57*(3), 436–437.

Burggräf, P., Steinberg, F., Becher, A., Sauer, C. R., & Müller, N. (2024). Optimize existing structures and workflows: Using a real-time locating system (RTLS) to enhance cleanliness in modern production areas. *Procedia CIRP*, *130*, 1494–1498.

Cao, H., Zhang, D., & Yi, S. (2023). Real-time machine learning-based fault detection, classification, and locating in large-scale solar energy-based systems: Digital twin simulation. *Solar Energy*, *251*, 77–85.

Cordero, D. A. (2025). Migration and the necessity of real-time locating systems in infection control. *Infection and Chemotherapy*, *57*(3), 438–439.

Dafhalla, A. K. Y., Isam, H. M., Ahmed, A. E. T., Elobaid, M. E., & Adam, T. (2025). Performance evaluation and QoS optimization of routing protocols in vehicular communication networks under delay-sensitive conditions. *Computers*, *14*(7), 285.

Danh, L. V. Q., Hoan, T. N. K., Pha, H. V., & Tram, N. T. (2023). An early drowning detection system capable of locating swimmers in real time. *Lecture Notes in Networks and Systems*, *485*, 175–186.

Dani, A., Rane, D., Bansal, S., & More, A. (2025). RFID-based automatic pathway for emergency vehicles. In *2025 IEEE International Students Conference on Electrical, Electronics and Computer Science (SCEECS 2025)*.

Deva Hema, D., & Rajeeth Jaison, T. (2024). Efficient collision risk prediction model for autonomous vehicle using novel optimized LSTM-based deep learning framework. *International Journal of Intelligent Transportation Systems Research*, *22*(2), 352–362.

Devi, G. D., Sudhir, S., & Taejashwar, R. B. (2025). Hand gesture-based emergency alert system for women's safety. In *2025 International Conference on Data Science and Business Systems (ICDSBS 2025)*.

Feng, S., Zhu, H., Wu, Y., Wang, J., & Jain, A. (2025). A dynamic capsule network-based physical alert monitoring system for elderly people with disabilities. *Engineering Applications of Artificial Intelligence*, *162*, 112556.

Ghafoorpoor Yazdi, P., Sbaragli, A., Peters, L., Pilati, F., & Thiede, S. (2024). Cyber physical system for reconfigurable learning factories: Combining 3D simulations, reconfigurable layouts and real-time locating systems. *Lecture Notes in Networks and Systems*, *1059*, 28–35.

Gu, H., Jiang, Y., Li, F., & Li, X. (2025). A numerical prediction model for the global collision behavior of marine risers. *Marine Structures*, *103*, 103817.

Hafid, Y. E., Ligabi, T., & Zahraoui, Y. (2025). Autonomous driving model with collision prediction for urban and extra-urban environments. *International Journal of Robotics and Control Systems*, *5*(1), 555–569.

Han, Y., Lin, X., Pan, D., Thomson, R., & Mizuno, K. (2024). Pedestrian trajectory prediction method based on the Social-LSTM model for vehicle collision. *Transportation Safety and Environment*, *6*(3), tdad044.

Hayes, N. W., Maldonado, B. P., Tang, M., Wang, P., & Hun, D. (2024). Optimization of prefabricated component installation using a real-time evaluator (RTE) connection locating system. In *Proceedings of the International Symposium on Automation and Robotics in Construction* (pp. 65–72).

Jonek-Kowalska, I., & Wolniak, R. (2025). *Exclusion in Smart Cities: Assessment and Strategy for Strengthening Resilience* (p. 224). Routledge.

Jurkus, R., Venskus, J., Markevičiūtė, J., & Treigys, P. (2025). Enhancing maritime safety: Estimating collision probabilities with trajectory prediction boundaries using deep learning models. *Sensors*, *25*(5), 1365.

Kim, M. H., Kim, J., Ra, H., Lee, H., & Kim, H. (2024). Identifying contact time required for secondary transmission of *Clostridioides difficile* infections by using real-time locating system. *Emerging Infectious Diseases*, *30*(5), 908–915.

Kim, M. H., & Park, Y. S. (2024). Use of a real-time locating system in infection control. *Infection and Chemotherapy*, *56*(4), 427–431.

Kowshik, M. R., Narayan, A., Nagarajan, P. M., & Gopalan, S. (2025). Dynamic dual-function safety wristband: A simulation-driven design for enhancing women's security and personal safety. *Procedia Computer Science*, *259*, 386–397.

Krugh, M., Waldherr, F., Küpper, C., Fürst, D., & Mears, L. (2024). Testing scenario for comparison of real-time locating system in automotive manufacturing. *Manufacturing Letters*, *41*, 1596–1606.

Latha, M., Fasiha, A. S., & Janusri, M. (2025). AI-driven video surveillance for safety using YOLO with IoT integration and intelligent alerts. In *2025 International Conference on Data Science and Business Systems (ICDSBS 2025)*.

Li, P., Xie, J., Ding, J., Wu, W., & Huang, G. Q. (2024). Real-time locating system-enabled digital twin for crane operation safety monitoring on construction sites. In *IEEE International Conference on Automation Science and Engineering* (pp. 2689–2696).

Magar, H., Ubale, A. S., Rajule, N., Gupta, R. R., & Philip, V. (2025). Case studies and real-world deployment examples of VANETs (Vehicular Ad-hoc Networks). *Studies in Computational Intelligence*, *1207*, 303–339.

Makhdoomi, P., Mardani, A., & Pashaei, M. H. (2025). A rapid surrogate model for collision risk prediction in real-time optimization of slender flexible manipulators with obstacle avoidance. *Iranian Journal of Science and Technology, Transactions of Mechanical Engineering*.

Mohsin, A. S. M., Choudhury, S. H., & Muyeed, M. A. (2025). Automatic priority analysis of emergency response systems using Internet of Things (IoT) and machine learning (ML). *Transportation Engineering*, *19*, 100304.

Monni, F., Dorazio, M., Bernardini, G., Lovreglio, R., & Quagliarini, E. (2025). Interactive wayfinding systems in emergency and evacuation conditions: Design of an integrated building component and CAVE-based assessment. In *2025 IEEE International Workshop on Metrology for Living Environment (MetroLivEnv 2025)* (pp. 362–366).

Nivishna, S., Sneha, R., Sunmathi, S., & Yakshini, M. (2025). Health centric driver safety and vehicle speed optimization system. In *Proceedings of the International Conference on Visual Analytics and Data Visualization (ICVADV 2025)* (pp. 1128–1133).

Ochoa-de-Eribe-Landaberea, A., Zamora-Cadenas, L., & Velez, I. (2024). Untethered ultra-wideband-based real-time locating system for road-worker safety. *Sensors*, *24*(8), 2391.

Odilova, G., Zaripova, M., Jabbarova, A., Allashev, A., & Akhrorova, M. (2025). Information security framework for online language education using differential privacy and secure multi-party computation algorithm. *Journal of Internet Services and Information Security, 15*(1), 96–106.

Panda, M., Ghosh, S., Mandal, S., Datta, S., & Roy, M. (2025). Real-time session monitoring application using machine learning algorithm. *Lecture Notes in Networks and Systems, 1342*, 279–291.

Park, H., Park, Y.-S., Park, S., Kim, D., & Chong, D.-Y. (2025). Prediction of ship damage severity in small-vessel collisions using a random forest model. *Ocean Engineering, 341*, 122644.

Patrick Sullivan, B., Yazdi, P. G., & Thiede, S. (2023). Total cost of ownership of real-time locating system (RTLS) technologies in factories. *Procedia CIRP, 120*, 822–827.

Pilati, F., Sbaragli, A., Ruppert, T., & Abonyi, J. (2024). Goal-oriented clustering algorithm to monitor the efficiency of logistic processes through real-time locating systems. *International Journal of Computer Integrated Manufacturing, 37*(10–11), 1359–1375.

Rosak-Szyrocka, J., & Wolniak, R. (2025a). *AI-driven Sustainability: The Future of Human Resources Management* (p. 224). Taylor & Francis Group.

Rosak-Szyrocka, J., & Wolniak, R. (2025b). *Smart Universities in Smart Cities: Shaping the Future of Education and Urban Innovation* (p. 232). Routledge.

Sayfoori, R., Huang, M.-H., Naderi, A., Frostig, R. D., & Cao, H. (2025). High-precision UWB sensor-based real-time locating system for rodent behavioral studies. *IEEE Sensors Journal, 25*(13), 25551–25559.

Shadpey, S., Aghamohammadi, M. R., & Sobbouhi, A. (2025). A novel SVM ensemble classifier for predicting potential blackouts under emergency condition using on-line transient operating variables. *Electric Power Systems Research, 241*, 111357.

Shamsollahi, D., Moselhi, O., & Khorasani, K. (2024). Data integration using deep learning and real-time locating system (RTLS) for automated construction progress monitoring and reporting. *Automation in Construction, 168*, 105778.

Sherif, F. F., & Ahmed, K. S. (2024). Medical equipment real-time locating system in hospitals based on bluetooth low energy. *Lecture Notes in Computer Science, 14848*, 142–151.

Sidiropoulos, A., Bechtsis, D., & Vlachos, D. (2025). Implementing an Industry 4.0 UWB-based real-time locating system for optimized tracking. *Applied Sciences, 15*(5), 2689.

Skotnicka-Zasadzień, B., & Wolniak, R. (2026). Artificial intelligence in public administration. In *Routledge Studies in Innovation, Organizations and Technology* (p. 170). Routledge.

Skubis, I., Wolniak, R., & Grebski, W. W. (2024). AI and human-centric approach in smart cities management: Case studies from Silesian and Lesser Poland Voivodships. *Sustainability, 16*, 18.

Sneha, S., Sriranjini, S., Thupakula, H., & Balaji, M. (2025). An IoT-enabled intelligent traffic management system with CNN-based accident detection and dynamic rerouting using K-means and Dijkstra's algorithm. In *Proceedings of the International Conference on Multi Agent Systems for Collaborative Intelligence (ICMSCI 2025)* (pp. 423–430).

Tao, M., Sun, X., & Han, P. (2025). A prediction model for coupled in-line and cross-flow vortex-induced vibration of a near-wall cylinder suffering collision. *Proceedings of the International Conference on Offshore Mechanics and Arctic Engineering (OMAE), 3, v003t06a017.*

Tebaldi, L., Volpi, A., & Azzolini, G. (2025). Real-time locating systems: A bibliometric analysis. In *Proceedings of the 2025 11th International Conference on Engineering Applied Sciences and Technology (ICEAST 2025)* (pp. 41–44).

Tundis, A., Hummel, M., Gunkel, J., & Savaglio, C. (2025). AI-based concepts for crisis propagation forecasting and early warning in urban areas. In *Proceedings of the 21st International Conference on Distributed Computing in Smart Systems and the Internet of Things (DCOSS-IoT 2025)* (pp. 805–811).

Valiollahi, S., Rodriguez, I., Zhang, W., Sharma, H., & Mogensen, P. (2024). Experimental evaluation and modeling of the accuracy of real-time locating systems for industrial use. *IEEE Access*, *12*, 75366–75383.

Vigram, K. M., Ramji, M., & Divyarupakala, D. (2025). Autonomous firefighting robot for enhanced fire detection and suppression. In *3rd International Conference on Intelligent Data Communication Technologies and Internet of Things (IDCIoT 2025)* (pp. 1070–1076).

Volpi, A., Montanari, R., Tebaldi, L., & Mambrioni, M. (2023a). Low-cost real-time locating system solution development and implementation in manufacturing industry. *Journal of Sensor and Actuator Networks*, *12*(4), 54.

Volpi, A., Tebaldi, L., Matrella, G., Montanari, R., & Bottani, E. (2023b). Low-cost UWB-based real-time locating system: Development, lab test, industrial implementation and economic assessment. *Sensors*, *23*(3), 1124.

von Wagner, M., Queck, A., Beekers, P., Boesing, J., & Chatterjea, S. (2024). Towards accurate and automatic emergency department workflow characterization using a real-time locating system. *International Journal of Healthcare Management*, *17*(2), 215–226.

Wang, Z., Liu, L., Cui, J., Zhao, D., & Li, M. (2023). MilliLoc: Human–computer interaction oriented acoustic millimeter-level real-time locating system. *Review of Scientific Instruments*, *94*(1), 015113.

Wolniak, R., Gajdzik, B., Grebski, M., Danel, R., & Grebski, W. (2024). Business models used in smart cities – Theoretical approach with examples of smart cities. *Smart Cities*, *7*, 4.

Wolniak, R., & Stecuła, K. (2024). Artificial intelligence in smart cities: Applications, barriers, and future directions—A review. *Smart Cities*, *7*, 3.

Yang, H., Li, X., & Pedrycz, W. (2023). Performance evaluation of three-dimensional UWB real-time locating auto-positioning system for fire rescue. *Intelligent Automation and Soft Computing*, *37*(3), 3039–3058.

Yuan, S., Li, Q., Lu, B., Liu, Y., & Gao, W. (2025). Trajectory prediction for fighter aircraft ground collision avoidance based on the model predictive control technique. *Aerospace Systems*, *8*(1), 61–70.

Zhang, S., Mao, R., Zhang, J., Xiao, L., & Cambria, E. (2025). MATADOR: Multimodal traffic accident prediction enhanced by multi-source aggregated emotion recognition. *Information Fusion*, *124*, 103335.

Zhang, Z., Wang, W., Wang, Q., Fu, M., & Guo, L. (2026). Research on performance test method of real time locating system. *Communications in Computer and Information Science*, *2417*, 312–320.

Zhong, J., Geng, F., Chen, Z., & Hou, W. (2025). Pedestrian collision safety performance prediction method based on deep learning models. *CMES: Computer Modeling in Engineering and Sciences*, *144*(1), 1–27.

Part III

Future of Safety

9 Ethical, Privacy, and Regulatory Considerations for AI Safety Systems

9.1 ALGORITHMIC ACCOUNTABILITY AND EXPLAINABILITY (XAI)

The growing reliance on algorithmic systems in industrial safety management has brought to the fore a structural tension between predictive power, operational autonomy, and organizational accountability. As artificial intelligence (AI) models assume evermore central functions in monitoring hazards, classifying anomalies, and supporting real-time decision-making, issues of how such systems reason become fundamental requirements rather than ancillary technical matters. Explainable AI is thus positioned as a strategic and epistemic framework oriented to reveal the inner logic, representational choices, and error profiles of safety-critical models. In industrial settings marked by multimodal data streams, complex human-machine interactions, and hierarchical decision lines, explainability assumes a dual role: it is at once an engineering discipline and a governance mechanism. It underpins not only transparency but also procedural legitimacy, ensuring traceability, justifiability, contestability, and, if necessary, overruling of algorithmic output by accountable human authorities.

A necessary part of algorithmic accountability is establishing clear pathways of responsibility. Often, predictive safety systems are based on high-dimensional inputs: sensor telemetry, machine usage logs, physiological signals from wearables, and environmental context variables. Every step in the pipeline, from data gathering to model inference, represents a design decision with associated ethical and regulatory ramifications. Accountability frameworks divide these decisions into discrete layers: data provenance, feature selection, model architecture, performance metrics, risk thresholds, and organizational protocols for model override. To the extent that these are well-documented, it becomes possible to map where bias arises, how uncertainty propagates, and which actors are authorized to intervene when predictions conflict with lived operational realities. In this sense, XAI is not just a descriptive tool but an institutional mechanism for rendering algorithmic behaviour answerable to the responsibility structure demanded by high-risk industrial work (Gerchick et al., 2025; Russell et al., 2023; Wieringa, 2023).

Technical literature distinguishes between global and local interpretability, and this difference is of particular importance in safety applications. Global explainability can be defined as how the model behaves over its full input space. It allows engineers to interpret the dominant patterns, recurrent dependencies, and sensitivity

DOI: 10.1201/9781003754213-12

profiles that show whether the system implicitly privileges some risk signals while downplaying others. A local explanation looks at the mechanisms underlying a single prediction. In industrial practice, these explanations become operational instruments, enabling supervisors to understand why an AI system classified a situation as hazardous, why the movement pattern of a worker was identified as anomalous, or why an automated alert was launched following a near-miss incident. Along with traceability requirements set by the EU Artificial Intelligence Act, both types of explainability contribute to facilitating organizations in demonstrating that model outputs are based on verifiable evidence rather than opaque statistical artefacts (Donia, 2025a; Nikidehaghani et al., 2023).

Another central dimension of XAI for safety systems concerns model calibration and the quantification of uncertainty. Predictive algorithms, usually trained on imbalanced datasets or in environments with non-stationary dynamics, have generally achieved high accuracy but are poorly reliable under rare or extreme scenarios. Safety-critical settings have no room to afford aggregate metrics such as the F1-score or ROC-AUC. Instead, they require calibrated confidence intervals, probabilistic risk thresholds, and ways of escalating ambiguous cases to human review. Explainability methods like SHAP, LIME, integrated gradients, and counterfactual analysis give organized insights into how uncertainty arises. When combined with uncertainty-aware architectures, such as Bayesian neural networks, deep ensembles, or conformal predictors, they allow organizations to view model output as an evidence layer within a broader socio-technical system rather than an autonomous authority. This shields organizations from overreliance on automated reasoning and reduces the risks of systemic error amplification (Schmidt et al., 2025; Valdivia et al., 2025).

The ethical dimension of algorithmic accountability cannot be divorced from the detection of bias. The industrial AI systems can accidentally learn correlations that encode structural inequalities. For example, classifiers of movement patterns may discriminate against older workers whose ergonomic profiles are different from the population on which the systems were trained, or algorithms for risk-scoring may systematically flag certain workers because of uneven sensor coverage. The XAI tools will let organizations identify whether protected attributes or contextually sensitive variables disproportionately drive predictions. Bias analysis, however, cannot be confined to its technical remediation. It requires organizational literacy as far as interpreting sociological drivers of imbalance is concerned and transparent communication strategies so that workers may understand how safety systems evaluate risk. Workers have increased trust in organizations where it is perceived that AI is applied uniformly and without functions for hidden surveillance. It is here that the adoption of AI-enabled safety platforms becomes sustainable (Horneber & Laumer, 2023; Metcalf et al., 2023).

The introduction of human-in-the-loop mechanisms strongly reinforces the normative grounding of explainability. Safety officers, production engineers, and first-line workers should have the ability to interrogate algorithmic reasoning in a manner appropriate for their professional roles. Explainability serves not only the data scientist who designs the system but also operators who experience risk directly. This means there have to be multilayered explanation modalities: visualizations for engineers, natural-language reasoning summaries for supervisors, and situationally grounded justifications for workers engaged in the task. Such multimodal interpretability underpins the feedback loop, where human expertise refines the model and it enhances human situational awareness. Such a process creates a dynamic accountability ecosystem in which responsibility is distributed but not diluted across human and computational agents (Trace & Hodges, 2024; Vavoula, 2024).

In this respect, ethical auditing acts as a complementary mechanism to technical explanation. Regular audits investigate whether the system behaves as expected, whether its performance drifts over time, and whether design assumptions remain valid under novel operational conditions. Algorithmic drift occurs when machinery ages in an industrial setting, worker routines evolve, environmental conditions change, or the maintenance schedule gets updated. XAI tools give auditors a forensic view of how internal model representations evolve with these changes. Audits also check for fairness in algorithms' decisions, the sufficiency of privacy safeguards, and the presence of unplanned implications related to surveillance. By creating periodic checkpoints, audits require accountability as an ongoing organizational responsibility rather than a one-time compliance exercise (Bartsch et al., 2024; Criado & Guevara-Gómez, 2024; Johnson, 2025; Martin, 2025).

Governance in algorithmic accountability needs clear institutional structures, marking decision rights and escalation pathways. According to the EU AI Act, high-risk AI systems must be supported by an auditable chain that links data governance policies, model documentation, technical validation procedures, and human oversight processes. This is where interpretable evidence by XAI outputs strengthens this chain for regulatory inspection, internal compliance reviews, and knowledge sharing across departments. The explainability outputs of safety-management teams are used, for example, when developing more refined interventions or justifying investment in infrastructure improvements. Using the same explainability outputs, data governance units validate whether privacy constraints or data minimization principles are satisfied without compromising safety objectives (Khan, 2024).

The Volkswagen safety reporting platform is the industrial case that combines these principles. Explainable AI identifies systemic risk hidden in large volumes of incident data anonymously provided by workers, thanks to a system that guarantees General Data Protection Regulation (GDPR) compliance through identity metadata segregation from risk-related data. This allows the safety managers to understand the rationale for risk clustering and effect factors and event co-occurrence patterns, thanks to XAI analysis, without disclosing personal data. This constitutes a step towards how algorithmic accountability can be combined with data protection requirements: explainability allows finding out the root causes of events, while privacy protection safeguards the integrity and trust of the reporting process. Workers believe this solution is dedicated to supporting safety improvements, not behavioural monitoring; hence, this generates high adoption, reinforcing the organizational safety culture.

Explanability also plays into incident reconstruction and post-event analysis. When an AI system feeds into a decision pathway – either by raising an alert, modifying a machine's operational profile, or adjusting risk levels – then organizations must reconstruct how the model evaluated the situation. It is here that XAI methods supporting structured reconstruction reports come in: determining which features weighed most in making the prediction, how model uncertainty changed over time, and if contextual anomalies played a role. These reports form the basis for internal investigations and external regulatory inquiries. They let organizations learn from incidents in a systematic way, turning individual events into collective knowledge assets.

Implementing XAI in the practice of industrial safety requires a socio-technical design philosophy. There is no automatic translation of technical transparency to responsible AI. The processes of explanation need anchorage in organizational routines such as toolbox meetings, shift handovers, near-miss debriefings, and cross-functional safety committees. Explainability becomes, in these contexts, a shared interpretative language that integrates engineering knowledge, worker experience, and managerial responsibility. In conjunction with robust accountability frameworks, explainability enables an organizational posture that approaches AI not as a source of opacity but as a catalyst for informed, evidence-based, and human-centred safety governance (Briscoe et al., 2025; Diakopoulos, 2025; Miller, 2025).

Table 9.1 articulates the multidimensional nature of algorithmic accountability in the form of how technical, organizational, and ethical dimensions have to come together in order to effectively regulate AI-driven safety systems at work. Each dimension – from responsibility mapping and traceability to human oversight and

TABLE 9.1
Algorithmic Accountability

Dimension	Core Questions	Practical Implications for Industrial AI Systems
Responsibility mapping	Who is responsible for data collection, model design, validation, and system oversight?	Clear assignment of duties to engineers, safety managers, data protection officers, and vendors; establishment of escalation pathways.
Traceability	Can all decisions, data flows, and model updates be reconstructed and audited?	Maintenance of detailed logs, dataset documentation, model versioning, and incident records.
Auditability	Do systems permit internal and external auditing?	Implementation of standardized documentation, access controls, and periodic algorithmic safety reviews.
Data governance	Is data used in a legal, necessary, and proportionate manner?	Enforcement of minimization, secure storage, retention limits, and pseudonymization protocols.
Performance validation	How is model accuracy, robustness, and error distribution evaluated?	Continuous performance monitoring, drift detection, and testing under realistic operational conditions.
Bias and fairness testing	Does the system produce systematically biased results?	Regular bias audits, dataset rebalancing, and fairness-driven model adjustments.
Human oversight	What are the mechanisms ensuring that humans can override, question, or contextualize model outputs?	Multilayered human-in-the-loop verification, clear guidance for override procedures, and worker training.

(Continued)

TABLE 9.1 (CONTINUED)

Dimension	Core Questions	Practical Implications for Industrial AI Systems
Explainability mechanisms	Can users comprehend why an AI system issued a certain alert or classification?	Deployment of XAI tools, conceptual explanations, and decision visualizations tailored to worker roles.
Incident accountability	How are algorithmic errors documented, reported, and remediated?	Formalized post-incident review processes, root-cause analysis, corrective actions, and traceable documentation.
Ethical compliance	Is system behaviour in line with ethical standards beyond legal requirements?	Institutionalization of ethical review boards, participatory consultations, and continuous governance evaluations.

Source: Author's own work.

ethical compliance – represents a particular domain in which an organization must clearly establish the rules, evidence-based processes, and verifiable records. Taken together, these dimensions suggest that accountability doesn't stop at model accuracy but extends into realms such as data governance, performance validation, mitigation of bias, auditability, and the ability of humans to interrogate or override algorithmic outputs. As the table below shows, practical implementation of these requirements turns AI from a black-box tool into a transparent and auditable part of the safety-management system, where predictive technologies strengthen rather than weaken worker trust, legal compliance, and organizational resilience.

9.2 GDPR, ISO 45001, AND THE AI ACT IN PRACTICE IN THE INDUSTRY

The intersection of data protection, safety management, and algorithmic regulation is one of the most daunting governance challenges facing industry today. As organizations increasingly deploy AI-based safety systems, their position within the coexistence of the GDPR, ISO 45001 occupational safety standards, and the forthcoming EU Artificial Intelligence Act presents a multilevel regulatory environment orienting organizational responsibilities at both the operational and strategic levels. Although each of these frameworks was separately designed from distinct institutional traditions – data protection law, occupational health and safety regulation, and AI – specific risk governance in industrial practice – they are in constant interaction. It is their combination that defines how companies capture data, construct safety protocols, validate AI models, and record compliance. The challenge facing industrial organizations is not to meet the requirements of each regime in isolation but to distil them into a coherent governance structure that will be resilient to legal challenge yet supportive of effective technological implementation.

The best starting point is to align the principle of data minimization under the GDPR with operational requirements for AI-driven safety monitoring. Safety systems will increasingly rely on high-frequency sensor data, video analytics, wearable telemetry, and behavioural logs. The GDPR does not prohibit such data processing; instead, it forces organizations to justify each data type via necessity analysis. That is, it demands that firms articulate, in a precise and documented way, why a given stream of data is necessary for the prevention of risk within the meaning of ISO 45001. This position requires employers to identify hazards, assess risks, and implement controls, and often, such AI systems support the process of identifying hidden anomalies or emerging threats. This relationship between the two frameworks becomes operational when Data Protection Impact Assessments are conducted alongside ISO 45001 hazard assessments. The point at which the two assessments interlink is where organizations demonstrate that data processing is a proportionate response to demonstrable risk, rather than some form of unwarranted surveillance. This alignment is thus a means of assuring legitimacy among workers while avoiding regulatory sanctions (Cuevas-Cárdenas et al., 2025; Dolgunyürek & Ada, 2025; Đurić et al., 2025; Shchipanov, 2025).

Actual industrial-scale implementation of GDPR in AI systems requires mechanisms of data governance beyond statements of compliance. In other words, this means determining retention periods and access rights. Strategies for pseudonymization or anonymization, particularly with the challenge of real-time monitoring of workers, may be complicated if anonymization impacts the ability to track longitudinal risk patterns, for example, in ergonomic assessment or near-miss detection. GDPR allows controlled pseudonymization on the condition that the reidentification key has to be kept securely stored and access strictly limited to only authorized personnel. Another important stipulation is that organizations have to make sure automated decisions cannot produce legally or materially significant consequences without human intervention. This crosses with requirements under ISO 45001, which explicitly covers worker participation in safety processes. Human-in-the-loop decision models accommodate both frameworks to ensure the AI systems do not autonomously take disciplinary measures or allocate blame; the function of the systems is to provide evidence to underpin safety deliberations carried out by competent personnel.

The new EU Artificial Intelligence Act introduces another layer with more requirements for industrial AI systems defined as 'high-risk'. Most safety-related applications of AI are considered high-risk, since they involve health, physical integrity, or workplace risks. As such, the Act requires organizations to put in place a quality management system documented for the full life cycle of AI: dataset curation, model development, validation procedures, incident logging, cybersecurity safeguards, and human oversight design. The other very important, unique feature of the AI Act is the technical documentation that will enable post-deployment traceability and scrutiny. That means companies will have to keep line-level logs of model behaviour, training datasets, updates of parameters, and performance drifts. These now tie in directly with ISO 45001 on maintaining records concerning incidents, hazard analysis, and corrective action. This in practice means teams have to embed AI documentation within existing occupational safety management documentation rather than having it as a free-standing artefact (Beltrán, 2025; Bignami et al., 2025; Jonek-Kowalska & Wolniak, 2025; Kotter et al., 2025; Potočnik & Fujs, 2025; Rosak-Szyrocka & Wolniak, 2025b; Weis et al., 2026; Wolniak et al., 2024).

The interaction of the AI Act with GDPR will be most pertinent in the industrial context, whenever biometric or behavioural data is dealt with. The AI Act further adds to the requirements without replacing the GDPR, where GDPR will focus on the legality of data collection, and the AI Act will focus on the safety, transparency, and reliability of AI systems processing such data. The kind of dual compliance structure that this reflects will most likely drive organizations towards system design that accommodates both lawful conditions for data processing and requirements of technical robustness. Consider an AI system set up for identifying hazardous behaviours through body posture analysis; an organization will have to justify the biometric treatment under the GDPR, establish strict controls for access, and at the same time meet the requirements under the AI Act regarding the quality of datasets, error rate testing, and bias mitigation. Such dual scrutiny will push industrial organizations towards more disciplined evidence-oriented development practices and reduce tolerance for ad hoc experimental deployments without comprehensive validation (Bharanitharan & Kaur, 2026; Campanile et al., 2026; Okonicha & Sadovykh, 2026; Wolniak et al., 2025).

ISO 45001 further stabilizes this regulatory ecosystem because it provides the underlying architecture of safety management on which AI-specific obligations can be mapped. It requires an organization to systematically identify hazards, consult workers, control risks, and evaluate safety performance. AI technologies become integrated into this system rather than serving as stand-alone monitoring tools. For example, in the case of using computer vision either for machine guarding or collision prevention, ISO 45001 would require consultations with workers around whether such technology could introduce psychosocial risks, ergonomic side effects, or even false-alarm stress. All of these have to be factored into the risk assessment, training protocols, and emergency plans. Further, the emphasis on continual improvement within the standard ensures that AI systems are periodically subjected to review not only for technical performance but also for impacts on the worker experience. Thus, ISO 45001 offers an institutional framework within which AI can be holistically assessed, weighing efficiency gains against human-centred concerns.

Another dimension arises in supply chains, in particular in large multinational industrial groups. Both the GDPR and the AI Act impose responsibilities not only on system users but also on developers, vendors, and integrators. In the near future, large industrial companies will increasingly rely on third-party suppliers of AI for risk analytics platforms, wearable safety devices, or predictive maintenance tools (Ivković et al., 2025; Luckner & Lauer, 2025; Yap, 2025). These arrangements will be managed within robust contractual frameworks that will detail an array of data protection duties, incident-reporting timelines, model update obligations, and audit rights. ISO 45001 reinforces this demand by insisting on control over externally provided processes. Suppliers will therefore need to be in a position to demonstrate that their AI systems comply with the GDPR and the AI Act, while large industrial companies need to be in a position to verify such claims. This will generally mean asking for model documentation, proof of dataset integrity, or certification against emergent AI conformity assessment schemes. The outcome is a more mature and transparent supply-chain governance model in which AI is subject to the same scrutiny as other critical safety equipment (Alibeigi, 2025; Beltrán, 2025; Martínez et al., 2025; Skotnicka-Zasadzień & Wolniak, 2026; Skubis et al., 2024).

While this interplay of the three regulatory pillars introduces new layers within organizational workflows, this does indeed have a truly transformational impact. A representative example could be incident reporting: whereas traditionally, safety incidents had to be logged manually either on paper-based forms or their digital equivalent, introducing AI-enabled platforms will mean adding new layers of documentation: algorithmic contributions to incident detection, model confidence levels, feature relevance explanations, and flags showing whether automated processing influenced human decision-making. Workers have a right to be so informed under the GDPR, while under the AI Act, such documentation has to be available to independent auditors and regulators (Ceccon et al., 2025; Diaz et al., 2025; Lawo et al., 2025; Westerstrand, 2025). ISO 45001 provides the context within which such reports feed into corrective and preventive actions. In other words, incident review committees will no longer have to make sense just of factual narratives but also of algorithmic outputs that integrate technical and human evidence into one unified assessment (Asbury, 2024; Ioana & Dumitraşcu, 2025; Olena et al., 2024; Podrecca et al., 2024).

The second pillar of pragmatic compliance is worker participation. GDPR entitles workers to know how their data are being processed, whereas ISO 45001 requires consultation on safety technologies. For its part, the AI Act goes one step further in identifying human oversight mechanisms and transparency appropriate to the context of use. These disparate requirements, taken together, mean that organizations need to design communication plans capable of rendering technical material into plain language (Balcı & Bozkurt, 2024; Górny, 2024). Firms may need to explain what information is collected, how long it's retained, what the conceptual functioning of an algorithm is, and how alerts will be interpreted by human supervisors in the specific context of deploying a wearable device able to detect motion patterns indicative of worker fatigue. Communication builds trust and reduces perceptions of surveillance, particularly in contexts where labour relations have been contentious. It enables procedural legitimacy for AI systems in a manner that makes workers aware of limitations, safeguards, and entitlements around automated monitoring (Sapienza & Palmirani, 2026; Tronnier et al., 2026).

Auditing practices are expanding due to regulatory convergence. The GDPR calls for periodic audits regarding data protection practices, while ISO 45001 mandates internal audits of safety-management systems. Added focus under the AI Act comes with model evaluation, robustness testing, bias assessment, and logging requirements. Combined audit models will thus be multilayered, from technical audits focused on model architecture and dataset quality to legal audits reviewing compliance with data protection, operational audits considering how the outputs will be integrated into workflows, and including safety audits in light of coherence with hazard-control procedures. Many industrial firms work towards integrated audit frameworks in which cross-functional teams assess systems from interdisciplinary perspectives. This reduces fragmentation and supports the creation of comprehensive evidence-based compliance documentation (Finck, 2025; Mahajan et al., 2025; Mandal et al., 2025; Marchiori et al., 2025; Schrepel, 2025; Skoczylas, 2025).

It is in the shaping of corporate culture that GDPR, ISO 45001, and the AI Act find their practical implementation. Compliance does not stop at some purely technical exercise but shapes the very conceptualization of responsibility, authority, and trust within those AI-mediated environments. Those companies that treat compliance as a rigid obligation rarely achieve meaningful adoption since workers view such technologies as imposed from the outside. On the other hand, organizations that integrate those regulations within a participatory continuous improvement culture tend to develop much more resilient and adaptive safety systems. These are those that treat data protection as part of dignity at work, interpret standards on safety as shared responsibilities, and see AI governance not as a constraint but as an instrument that strengthens organizational integrity (Gioiosa et al., 2025; Kapitsaki & Papoutsoglou, 2025; Pandit, 2026).

Table 9.2 summarizes in tabular format how the AI Act translates into concrete operational obligations in an industrial setting and underlines how abstract regulatory principles become actionable organizational practices. From high-risk classification

TABLE 9.2
AI Act in Industry Practice

AI Act Requirement	Practical Implications for Industrial Deployment
High-risk system classification	Compulsory conformity assessment before deployment and periodic reassessment during operation.
Quality management system	Implementing documented processes for data handling, model validation, incident logging, and oversight.
Data governance requirements	Verification of dataset representativeness, accuracy, and freedom from structural bias.
Technical documentation	Creation of comprehensive model files, training-data descriptions, performance metrics, and safety validations.
Logging obligations	Continuously log model activity, errors, updates, anomalies, and human overrides.
Transparency obligations	The system should clearly communicate to users its purpose, limitations, and operational boundaries.
Human oversight mechanisms	Override protocols, training, and supervision guidelines shall be provided by the management.
Accuracy and robustness standards	Regular testing in realistic operational settings, including stress scenarios and edge cases.
Cybersecurity measures	Protecting training data, inference pipelines, and model parameters from tampering and adversarial attacks.
Post-market monitoring	Collection of operational evidence, performance drift analysis, and reporting of safety incidents.
Conformity assessment	Third-party or internal evaluation that validates conformance to technical or legal requirements.
Risk-management system	Continuous identification, mitigation, and documentation of the risks related to AI operation.
User training requirements	Training of employees and supervisors in system functionality, safe use, and recognition of failure.
Duty to update documentation	Ongoing revision of technical files and risk analyses in the event of model retraining or modification.
Interaction with other regulations	Integration of the tasks set by the AI Act with GDPR, ISO 45001, product-safety rules, and sector-specific regulations.

Source: Author's own work.

over quality management systems, transparency duties, and cybersecurity safeguards, each requirement imposes specific procedures on companies to be implemented at different points in time during the AI life cycle. Entries illustrate very well that compliance goes much further than initial system validation; it requires continuous monitoring, rigorous documentation, systematic risk management, and structured human oversight embedded in day-to-day operations. Furthermore, the table underlines the fact that the AI Act does not operate in isolation; its provisions need to be integrated with existing frameworks such as the GDPR and ISO 45001. All these elements put together show that integrating the AI Act within industrial practice requires disciplined governance, interdisciplinary coordination, and sustained organizational commitment to responsible and safe AI deployment.

9.3 PARTICIPATORY PRIVACY-PRESERVING SAFETY SYSTEM DESIGN

AI-driven safety systems form part of new modes of interaction between workers, digital infrastructures, and organizational decision-making. So much conventional safety engineering was strongly top-down in its logic: experts designed control mechanisms, and workers were simply recipients of the resulting rules and procedures. AI-driven systems blur this arrangement because their efficacy depends as much on technical accuracy as on social legitimacy, contextual knowledge, and the willingness of the workforce to engage with data and interfaces informing risk interpretation. The emphasis in the development of this approach to safety systems avoids erosion of autonomy when the use of monitoring technologies increases, making sure that no safety improvements occur at the cost of psychological comfort, dignity, or even trust.

Participatory design in system development begins at the very beginning of the conception process. Rather than define system requirements through expert analysis alone, an organization brings a representative set of workers into workshops, scenario-modelling exercises, and hazard-mapping sessions. Those interactions reveal knowledge not captured in formal data sets, such as practices used to manage unexpected machine behaviour, informal coordination signals between team members, and situational cues associated with risks that are not documented in standard procedures. This informs system architecture by ensuring that data sources, sensor placement, user interfaces, and alert thresholds reflect real-world experience. The workers also identify where data collection would be unnecessary or intrusive. Given that privacy preservation depends on selecting the minimum set of data needed to achieve safety objectives, participatory sessions often result in the removal of sensors with little value or those that could create perceptions of surveillance (Li et al., 2024; Lin, 2024; Verma et al., 2024).

Designing privacy-preserving systems starts with implementing purpose-bound data flows. While the limitation of purpose is certainly one cornerstone of European data protection law, it relies in practice on social negotiation to define the exact boundaries between legitimate safety monitoring and unacceptable behaviour profiling. Participatory processes support boundary setting if workers express their concerns about data misuse, power imbalance, or possible disciplinary consequences.

Such discussions drive the architecture of access controls, retention policies, and data handling procedures. Workers may, for instance, agree on the use of aggregated heat maps for hazardous traffic flow detection, while they refuse to allow monitoring of individually identifiable trajectories of movement. Consequently, the translation of these normative positions into technical constraints – such as local model computation on edge devices, rigorous anonymization rules, or multilevel access rights – will always result in systems that fulfil safety requirements and meet privacy expectations (Fadhil et al., 2024; Stanculescu et al., 2024).

Participatory mechanisms influence the design of both algorithmic outputs and interaction modalities. Indeed, AI-driven safety systems often produce complex risk metrics or predictive indicators that are hardly interpretable by non-experts. Workers need to understand how alerts pertain to real operations, how to react to them, and how to contest them when they seem off target. Co-design workshops thus become a forum for the exploration of alternative explanation formats: simple, causal stories; colour-coded interfaces; temporal trend visualizations; or context-dependent guidance. When workers take part in the choice of the form of explanation, they become more inclined to embed AI's insights into their daily routines. What matters here is not that workers should understand the underlying mathematics of machine learning; the approach focuses instead on devising communicative channels that fit their cognitive models and work rhythms. In this way, explainability becomes a relational property of the human-machine system rather than a merely technical feature (Martinez-Bejar et al., 2023; Rosak-Szyrocka & Wolniak, 2025a; Tamilselvi, 2024; Wolniak & Stecuła, 2024).

Perhaps most importantly, participatory methods serve to rectify power asymmetries. AI systems sometimes displace managerial authority by rendering certain forms of knowledge more salient at the expense of others. Statistical models, for instance, often give an advantage to quantifiable metrics at the expense of qualitative judgement acquired over a decade or more of practice. Participatory processes formally introduce experiential knowledge into the design process. Workers become active coproducers of epistemic authority rather than passive subjects of data extraction. This, in turn, minimizes resistance to technological change and extends a sense of collective ownership. In industrial settings with strong traditions of union representation, such an approach also positions AI adoption in line with established principles of social dialogue, thus mitigating adversarial dynamics that commonly materialize at the intersection between automation and labour relations (Chaymae et al., 2025; Zhang et al., 2025).

Participation plays an important role in the identification of these adverse, often unintended, consequences. Well-intentioned safety systems can engender psychosocial risks, including increased stress associated with continuous monitoring, reduced sense of autonomy, or reinforcement of hierarchical barriers when algorithmic outputs are used in assessing individual performance. Participation provides an early warning system for these risks. At the design phase, workers can highlight points in their workflow where surveillance may be felt particularly acutely, or indeed, where algorithmic alerts may interfere with concentration. Input like this flows into technical decisions related to data granularity, sampling frequency, user-interface timing, or the application of differential privacy techniques. It feeds into policy decisions at the organizational level, such as explicit guarantees that AI-generated alerts will never be used for disciplinary purposes or the development of review procedures whereby workers can contest algorithmic assessments (Li et al., 2024; Losavio, 2025).

Participatory design provides essential added value to privacy-preserving mechanisms because strong protection of privacy requires insight into how workers interpret risk, dignity, and personal space. Privacy is not just a category of law; it is also a socially constructed experience that differs by departments, jobs, and cultural backgrounds. Participatory design thus allows the contextualization of privacy measures. Intrusive monitoring can be unacceptable in settings where independence and high discretion are valued by workers, even if those were legally admissible. Workers may accept certain forms of monitoring in situations where collaboration and mutual visibility are at the heart of safety culture, providing clear benefits. This calls for flexible system design with, among others, secure multi-party computation, encrypted sensor networks, on-device processing, or selective obfuscation tailored to the social profile of the workplace (Dziurakh et al., 2024; Verma et al., 2024).

Participatory approaches lie at the heart of the development of feedback loops for system improvement. AI-driven safety systems are in continuous flux because of the accumulation of new data, retraining models, and shifting operating contexts. Without mechanisms for worker feedback, updates to systems risk drifting away from real operational needs. Participatory governance inserts regular consultation cycles whereby workers observe system behaviour, flag false alarms, suggest refinements, and comment on emerging hazards. This is a recursive process consistent with principles of continuous improvement in safety management, but extended here to include the algorithmic elements. Participation secures that AI systems remain adaptive, not just to data but also to evolving social dynamics in the workplace (Alipour et al., 2023; Lin, 2024; Yang et al., 2024).

Participatory design enhances the epistemic transparency of AI systems by considering the way risk is conceptualized within models and inviting workers to reflect on such conceptualizations. They thus develop an understanding of the assumptions that are entailed by system logics. It's not technical training in machine learning but rather the development of a shared conceptual vocabulary concerning risk factors, thresholds, and classification principles. Shared understanding allows workers to contextualize system outputs and guard against overreliance on algorithmic indicators. It also limits the risk of 'automation bias', whereby individuals uncritically defer to digital systems. In this way, the process of participatory design protects workers from what might well become the overreach of algorithms or underestimation of their own expertise by maintaining human interpretive authority.

Participatory, privacy-preserving design also has implications for organizational accountability structures. When workers have a say in the design, governance responsibilities become explicit: they know who is in control of the data, who is interpreting the alerts, and who has the authority to pause or override automated functions. And this clarity reduces the ambiguity and mystery that often surround AI deployment. From a governance point of view, participation helps embed safety systems in the organization's processes: risk committees, safety audits, human resources policies, and technical maintenance plans. It establishes channels for raising concerns, objections, or modification requests about the system. Such channels constitute checks on possible misuse of either data or technology. Rather than relying on compliance officers or external auditors, organizations develop internal oversight capacity distributed across multiple levels of the workforce.

The participatory approach underpins long-term organizational resilience. Industrial ecosystems operate in environments characterized by uncertainty, complex interdependencies, and continuously changing regulatory requirements. AI systems co-designed with workers are more robust since they embed diversity of perspectives on risk, operational context, and ethical constraints. Participation results in systems that are adaptive, transparent, and socially grounded. It reinforces trust, a key ingredient when AI systems intervene in safety-critical domains. Workers who trust the system will be more likely to wear devices consistently, report anomalies, and collaborate on AI-generated insights. In turn, trust reinforces data quality and improves model performance. This recursive dynamic is indicative of how participatory privacy-preserving design is not a soft complement to technical innovation but instead forms a structural precondition for sustainable AI-enabled safety systems (de Lima, 2023; Li et al., 2023; Yang et al., 2024).

BIBLIOGRAPHY

Alibeigi, A. (2025). Bridging the gap: Assessing India's Digital Personal Data Protection Act in light of the EU GDPR. *SN Computer Science*, *6*(7), 855.

Alipour, J., Mehdipour, Y., Karimi, A., Khorashadizadeh, M., & Akbarpour, M. (2023). Security, confidentiality, privacy and patient safety in the hospital information systems from the users' perspective: A cross-sectional study. *International Journal of Medical Informatics*, *175*, 105066.

Asbury, S. (2024). Navigating ISO 45001: Techniques, implementation tools and case studies. *Navigating ISO 45001: Techniques, Implementation Tools and Case Studies* (pp. 1–309). Routledge.

Balcı, R., & Bozkurt, Y. (2024). A new approach to examining the current state of ISO 45001 in the education sector: Gap analysis. *Journal of Environmental Health*, *87*(3), 26–31.

Bartsch, S. C., Milani, V., Adam, M., & Benlian, A. (2024). Algorithmic accountability: What does it mean for AI developers and how does it affect AI development projects. In *Proceedings of the Annual Hawaii International Conference on System Sciences* (pp. 5826–5835).

Beltrán, M. (2025). AI algorithms under scrutiny: GDPR, DSA, AI Act and CRA as pillars for algorithmic security and privacy in the European Union. *Computers and Security*, *158*, 104628.

Bharanitharan, K., & Kaur, G. (2026). A comparative analysis of data minimization principles: Evaluating GDPR and India's DPDP Act 2023. In *Lecture Notes in Networks and Systems, 1385 LNNS* (pp. 41–55). Springer.

Bignami, E., Darhour, L. J., Buhre, W., Cecconi, M., & Bellini, V. (2025). Artificial intelligence in healthcare: Tailoring education to meet EU AI-Act standards. *Health Policy and Technology*, *14*(6), 101078.

Briscoe, J., Kepler, G., Deford, D., & Gebremedhin, A. (2025). Algorithmic accountability in small data: Sample-size-induced bias within classification metrics. *Proceedings of Machine Learning Research*, *258*, 4915–4923.

Campanile, L., Iacono, M., Mastroianni, M., Riccio, C., & Viscardi, B. (2026). A TOPSIS-based approach to evaluate alternative solutions for GDPR-compliant smart-city services implementation. In *Lecture Notes in Computer Science, 15893 LNCS* (pp. 303–316). Springer.

Ceccon, M., Cornacchia, G., Dalle Pezze, D., Fabris, A., & Susto, G. A. (2025). Underrepresentation, label bias, and proxies: Towards data bias profiles for the EU AI Act and beyond. *Expert Systems with Applications*, *292*, 128266.

Chaymae, M., Youssef, G., & Saida, E. M. (2025). Systematic review for attack tactics, privacy, and safety models in big data systems. *Indonesian Journal of Electrical Engineering and Computer Science*, *37*(2), 1234–1250.

Criado, J. I., & Guevara-Gómez, A. (2024). Who evaluates the algorithms? An overview of the algorithmic accountability ecosystem. In *ACM International Conference Proceeding Series* (pp. 19–28).

Cuevas-Cárdenas, Z., Rodriguez-Mendoza, J., Evangelista-Guía, W., Zevallos-Vera, E., & Salazar-Sandoval, F. (2025). Reduction of occupational risks through the application of an SGSST based on the ISO 45001:2018 standard in the company Forli S.A.C | Reducción de riesgos laborales mediante la aplicación de un SGSST basado en la norma ISO 45001:2018 en la empresa Forli S.A.C. In *Proceedings of the LACCEI International Multi-Conference for Engineering Education and Technology (2025).*

de Lima, I. J. (2023). Safety, security and privacy in future airport terminals: A system theory perspective. *Transportation Research Procedia*, *72*, 2503–2509.

Diakopoulos, N. (2025). Prospective algorithmic accountability and the role of the news media. In *Philosophy of Engineering and Technology* (Vol. 44, pp. 125–143). Springer.

Diaz, J. Z., Brennan, T., & Corcoran, P. (2025). Navigating the EU AI Act: Foreseeable challenges in qualifying deep learning–based automated inspections of Class III medical devices. *International Journal of Advanced Manufacturing Technology*, *140*(11–12), 5869–5883.

Dolgunyürek, O., & Ada, S. (2025). Evaluation of the effectiveness of implementation of ISO 45001 occupational health and safety-management system in educational institutions | ISO 45001 iş sağlığı ve güvenliği yönetim sistemi'nin eğitim kurumlarında uygulanmasının etkinliğinin değerlendirilmesi. *Milli Eğitim*, *54*(246), 1197–1220.

Donia, J. (2025a). Algorithmic accountabilities and health systems: A review and sociomaterial approach. *Big Data and Society*, *12*(2).

Donia, J. (2025b). Algorithmic accountability. In *Understanding Accountability: New Perspectives on a Fractured World* (pp. 173–190). Springer.

Đurić, Ž., Cvijanović, N., Kostić, D., Vuković, J., & Vasiljević, N. (2025). An overview of certified ISO 45001 OH&S systems in the regional context. *Engineering Proceedings*, *99*(1), 2.

Dziurakh, Y., Kulyniak, I., Sarkisian, H., Chepil, B., & Vaskovych, K. (2024). Intrusion detection systems for smart tourism platforms: Safeguarding food safety and user privacy. *Journal of Internet Services and Information Security*, *14*(4), 484–498.

Fadhil, A. M., Din, N. M., Aripin, N. B. M., & Abed, A. A. (2024). Advanced privacy scheme to improve road safety in smart transportation systems. *International Journal of Computer Networks and Communications*, *16*(2), 71–86.

Finck, M. (2025). In search of the lost research exemption: Reflections on the AI Act. *GRUR International*, *74*(10), 903–904.

Gerchick, M. K., Encarnación, R., Tanigawa-Lau, C., Gutiérrez, A., & Metaxa, D. (2025). Auditing the audits: Lessons for algorithmic accountability from Local Law 144's bias audits. In *Proceedings of the 2025 ACM Conference on Fairness, Accountability and Transparency* (pp. 29–44).

Gioiosa, S., Chiavarini, B., D'Antonio, M., Cestari, M., & Rossi, E. (2025). A GDPR-compliant solution for analysis of large-scale genomics datasets on HPC cloud infrastructure. *Journal of Big Data*, *12*(1), 31.

Górny, A. (2024). Guidelines and needs for the implementation of the ISO 45001 requirements for shaping of safety in Industry 4.0. In *Lecture Notes in Mechanical Engineering* (pp. 107–123). Springer.

Horneber, D., & Laumer, S. (2023). Algorithmic accountability. *Business and Information Systems Engineering*, *65*(6), 723–730.

Ioana, A.-C., & Dumitraşcu, A.-E. (2025). Implementation of integrated management systems: ISO 9001, ISO 45001 and ISO 13485. In *Lecture Notes in Mechanical Engineering* (pp. 290–297). Springer.

Ivković, R., Vučković, D., Ječmenić, M., Petković, M., & Janković, N. (2025). The EU AI Act and its contractual and educational implications: AI, education, and the law of obligations. *International Journal of Cognitive Research in Science, Engineering and Education*, *13*(2), 123–145.

Johnson, D. G. (2025). Constituting algorithmic accountability. In *Philosophy of Engineering and Technology* (Vol. 44, pp. 11–28). Springer.

Jonek-Kowalska, I., & Wolniak, R. (2025). *Exclusion in Smart Cities: Assessment and Strategy for Strengthening Resilience* (p. 224). Routledge.

Kapitsaki, G. M., & Papoutsoglou, M. (2025). Evolution of repositories and privacy laws: Commit activities in the GDPR and CCPA era. *Journal of Systems and Software*, *230*, 112515.

Khan, M. (2024). Framing online speech governance as an algorithmic accountability issue. *Indiana Law Journal*, *99*(5), 37–61.

Kotter, E., D'Antonoli, T. A., Cuocolo, R., Shelmerdine, S. C., & Bellemo, A. (2025). Guiding AI in radiology: ESR's recommendations for effective implementation of the European AI Act. *Insights into Imaging*, *16*(1), 33.

Kyi, L., Santos, C., Shivakumar, S. A., Roesner, F., & Biega, A. (2026). Turning to online forums for legal information: A case study of GDPR's legitimate interests. In *Lecture Notes in Computer Science, 16183 LNCS* (pp. 153–177). Springer.

Lawo, D., Neifer, T., Roßbach, Y. A., & Boden, A. (2025). Solutions and challenges of AI practitioners in AI Act compliance. In *Proceedings of the 12th International Conference on Communities and Technologies* (pp. 1–11).

Li, B., Xu, W., Chen, J., Bai, G., & Yuan, X. (2024). LAMPS'24: ACM CCS workshop on large AI systems and models with privacy and safety analysis. In *Proceedings of the 2024 ACM SIGSAC Conference on Computer and Communications Security* (pp. 4888–4889). ACM.

Li, J., Hou, N., Zhang, G., Liu, Y., & Gao, X. (2023). Efficient conditional privacy-preserving authentication scheme for safety warning system in edge-assisted Internet of Things. *Mathematics*, *11*(18), 3869.

Lin, X.-J. (2024). On the unforgeability of privacy-preserving aggregation-authentication scheme for safety warning system in fog-cloud based VANET. *IEEE Transactions on Information Forensics and Security*, *19*, 10371–10372.

Losavio, M. M. (2025). Security and privacy in ambient intelligence, the Internet of Things and pervasive systems: Interrelationships for systems of public safety. In *Lecture Notes in Networks and Systems, 1285 LNNS* (pp. 678–687). Springer.

Luckner, S., & Lauer, W. (2025). Regulatory classification of AI-enabled products for medical use on the basis of the EU AI Act and MDR/IVDR | Regulatorische Einordnung KI-basierter Produkte für die medizinische Anwendung auf Basis von EU AI Act und MDR/IVDR. *Bundesgesundheitsblatt – Gesundheitsforschung – Gesundheitsschutz*, *68*(8), 854–861.

Mahajan, S., Agarwal, R., & Gupta, M. (2025). Algorithmic bias under the EU AI Act: Compliance risk, capital strain, and pricing distortions in life and health insurance underwriting. *Risks*, *13*(9), 160.

Mandal, S., Bhardwaj, S., Ghiyal, R., & Saju, S. J. (2025). Merging artificial intelligence with the existing legal frameworks: A comparative analysis of the EU's AI Act. In *Artificial Intelligence in Legal Systems: Bridging Law and Technology Through AI* (pp. 100–117). Springer.

Marchiori, S., Hopster, J. K. G., Puzio, A., Viehoff, J., & Frank, L. E. (2025). A social disruptiveness-based approach to AI governance: Complementing the risk-based approach of the AI Act. *Science and Engineering Ethics*, *31*(5), 25.

Martin, K. (2025). Johnson's algorithmic accountability and corporate accountability dissonance. In *Philosophy of Engineering and Technology* (Vol. 44, pp. 51–65). Springer.

Martínez, D., Molero, A., Calle, E., Ametller, D. C., & Jové, A. (2025). Large-scale web tracking and cookie compliance: Evaluating one million websites under GDPR with AI categorization. *Journal of Network and Computer Applications*, *242*, 104222.

Martinez-Bejar, R., García-González, M. S., Scalera, L., & Gasparetto, A. (2023). A modelling approach to privacy and safety issues in cyber-physical systems. In *Exploring Cyber Criminals and Data Privacy Measures* (pp. 241–252). Springer.

Metcalf, J., Singh, R., Moss, E., Tafesse, E., & Watkins, E. A. (2023). Taking algorithms to courts: A relational approach to algorithmic accountability. *ACM International Conference Proceeding Series* (pp. 1450–1462).

Miller, K. W. (2025). Algorithmic accountability gets the Deborah Johnson treatment. In *Philosophy of Engineering and Technology* (Vol. 44, pp. 29–50). Springer.

Nikidehaghani, M., Andrew, J., & Cortese, C. (2023). Algorithmic accountability: Robodebt and the making of welfare cheats. *Accounting, Auditing and Accountability Journal*, *36*(2), 677–711.

Okonicha, O., & Sadovykh, A. (2026). Enhancing GDPR compliance through NLP: Automated policy evaluation and legal document analysis. In *Lecture Notes in Networks and Systems, 1471 LNNS* (pp. 408–423). Springer.

Olena, C., Lina, F., Maxym, Y., Denys, Z., & Valerii, K. (2024). Assessment of the quality of the labour safety management system in accordance with the requirements of the international standard ISO 45001:2018 at the stage of operation. *Innovative Technologies and Scientific Solutions for Industries*, *2024*(1), 226–235.

Pandit, H. J. (2026). Simple now, complex later: The questionable efficacy of diluting GDPR Article 30(5). In *Lecture Notes in Computer Science, 16183 LNCS* (pp. 127–149). Springer.

Podrecca, M., Molinaro, M., Sartor, M., & Orzes, G. (2024). The impact of ISO 45001 on firms' performance: An empirical analysis. *Corporate Social Responsibility and Environmental Management*, *31*(5), 4581–4595.

Potočnik, J., & Fujs, D. (2025). Navigating uncharted waters: Select practical considerations in radiology AI compliance with the EU AI Act. *NPJ Digital Medicine*, *8*(1), 630.

Rosak-Szyrocka, J., & Wolniak, R. (2025a). *AI-Driven Sustainability: The Future of Human Resources Management* (p. 224). Taylor & Francis Group.

Rosak-Szyrocka, J., & Wolniak, R. (2025b). *Smart Universities in Smart Cities: Shaping the Future of Education and Urban Innovation* (p. 232). Routledge.

Russell, A. M., Bergman, B. G., Colditz, J. B., & Massey, P. M. (2023). Algorithmic accountability on social media platforms in the context of alcohol-related health behavior change. *Addiction*, *118*(1), 189–190.

Sapienza, S., & Palmirani, M. (2026). Algorithmic knowability: A unified approach to explanations in the AI Act. In *Communications in Computer and Information Science, 2576 CCIS* (pp. 185–209). Springer.

Schmidt, J.-H., Bartsch, S., Zweidinger, Y., & Benlian, A. (2025). Algorithmic accountability as a virtue or a mechanism? The ethical divide among AI developers. In *Proceedings of the Annual Hawaii International Conference on System Sciences* (pp. 677–686).

Schrepel, T. (2025). Decoding the AI Act: Implications for competition law and market dynamics. *Journal of Competition Law and Economics*, *21*(3), 374–389.

Shchipanov, A. V. (2025). Integration of occupational health and safety management systems with medical support based on GOST R ISO 45001—2020 and GOST R 59240—2020. *Bezopasnost Truda v Promyshlennosti*, *2025*(8), 80–87.

Skoczylas, D. (2025). Artificial intelligence algorithms in the context of cybersecurity: AI Act and the national cybersecurity system – opportunities and challenges for social and economic development in the Republic of Poland. In *Use of Algorithms in Criminal Law: A Central European Perspective* (pp. 61–73). Springer.

Skotnicka-Zasadzień, B., & Wolniak, R. (2026). Artificial intelligence in public administration. In *Routledge Studies in Innovation, Organizations and Technology* (p. 170). Routledge.

Skubis, I., Wolniak, R., & Grebski, W. W. (2024). AI and human-centric approach in smart cities management: Case studies from Silesian and Lesser Poland Voivodships. *Sustainability*, *16*, Article 18.

Stanculescu, D. A., Solomon, A., Chiaga, D., Siminski, J., & Merz, K. (2024). Advancing satellite safety through AI/ML, multi-agent systems and privacy enhancement techniques. In *Proceedings of the International Astronautical Congress (IAC), 3* (pp. 1991–2000).

Tamilselvi, M. (2024). A novel privacy enabled human health and safety monitoring system using IoT with smart wearable device. In *2nd International Conference on Self-Sustainable Artificial Intelligence Systems (ICSSAS 2024) Proceedings* (pp. 1021–1027).

Trace, C. B., & Hodges, J. A. (2024). The role of paradata in algorithmic accountability. *Knowledge Management and Organizational Learning*, *13*, 197–213.

Tronnier, F., Löbner, S., Lacombe, M.-H., & Rannenberg, K. (2026). Regulatory challenges in cybersecurity – A critical analysis of the EU AI Act. In *IFIP Advances in Information and Communication Technology, 742 IFIPAICT* (pp. 80–93). Springer.

Valdivia, A., Hyde-Vaamonde, C., & García Marcos, J. (2025). Judging the algorithm: Algorithmic accountability on the risk assessment tool for intimate partner violence in the Basque Country. *AI and Society*, *40*(4), 2633–2650.

Vavoula, N. (2024). Algorithmic accountability through the "human over the loop" in interoperable and EU AI-reliant large-scale IT systems for migration and security. *European Papers: A Journal on Law and Integration*, *9*(3), 1228–1249.

Verma, G. K., Wani, N. M., Rana, S., Kumar, N., & Tiwari, A. (2024). Dynamic-anonymous privacy-preserving authenticated aggregation for safety-warning system for the Internet of Vehicles. *Journal of Information Security and Applications*, *84*, 103830.

Weis, S., Montsch, C., Delpechithrage, T., & Nguyen, B. A. P. (2026). From regulation to implementation: Understanding the impact of the EU AI Act on public sector institutions in Germany. In *Lecture Notes in Computer Science, 15978 LNCS* (pp. 87–101). Springer.

Westerstrand, S. (2025). Fairness in AI systems development: EU AI Act compliance and beyond. *Information and Software Technology*, *187*, 107864.

Wieringa, M. (2023). Hey SyRI, tell me about algorithmic accountability: Lessons from a landmark case. *Data and Policy*, *5*, e2.

Wolniak, R., Gajdzik, B., Grebski, M., Danel, R., & Grebski, W. (2024). Business models used in smart cities – Theoretical approach with examples of smart cities. *Smart Cities*, *7*, 4.

Wolniak, R., Kowalska-Styczeń, A., Jonek-Kowalska, I., & Michalak, A. (2025). Methods for measuring the effectiveness of human-AI collaboration in an organization. In L. M. Camarinha-Matos, A. Ortiz, X. Boucher, & A. L. Soares (Eds.), *Hybrid Human-AI Collaborative Networks. 26th IFIP WG 5.5 SOCOLNET Working Conference on Virtual Enterprises, PRO-VE 2025, Porto, Portugal, October 27–29, 2025, Proceedings, Part II.*

Wolniak, R., & Stecuła, K. (2024). Artificial intelligence in smart cities: Applications, barriers, and future directions—A review. *Smart Cities*, *7*, 3.

Yang, Y., Zhang, L., Zhao, Y., Choo, K.-K. R., & Zhang, Y. (2022). Privacy-preserving aggregation-authentication scheme for safety warning system in fog-cloud based VANET. *IEEE Transactions on Information Forensics and Security*, *17*, 317–331.

Yang, Y., Zhang, L., Zhao, Y., Choo, K.-K. R., & Zhang, Y. (2024). Rebuttal to "On the unforgeability of 'privacy-preserving aggregation-authentication scheme for safety warning system in fog-cloud based VANET.'" *IEEE Transactions on Information Forensics and Security*, *19*, 10373–10374.

Yap, S. N. (2025). AI governance for sustainable development: Lessons for Malaysia from the EU AI Act. In *Integrating Legal Compliance with Sustainable Development Goals* (pp. 327–354). Springer.

Zhang, N., Ward, B., Clark, A., Zhao, Z., & Xiong, A. (2025). Introduction to the special issue on security and privacy in safety-critical cyber-physical systems. *ACM Transactions on Cyber-Physical Systems*, *9*(3), 25.

10 Workplace Safety in Industry 5.0

Customization, Collaboration, and Sustainability

10.1 EMOTIONAL AI AND WORKER-CENTRED DESIGN

Emotional artificial intelligence (emotional AI) can be defined as the use of artificial intelligence to understand and react to human emotion in digital communication between humans and machines. Emotional AI is generally seen as a subfield of affective computing, an umbrella term proposed by Picard (2000) that refers to the study and development of systems capable of processing emotional data in order to facilitate human-technology interaction and communication. In this context, emotional AI mostly concerns the automatic recognition of emotions especially those that can be inferred from facial expressions, vocal prosody and biometric signals, and the system's in-the-moment sensitivities to them (Calvo & D'Mello, 2010; McStay, 2018).

In order to better analyse how emotion-aware intelligent systems (EAISs) interpret and transform emotional user's state, we can identify the most important sources from which they get information. In the literature (Calvo & D'Mello, 2010; McStay, 2018; Picard, 2000), these channels are described by modalities, indicating different sources of signals employed for emotion recognition. The modalities that are commonly used in emotional AI and their relevancy towards human work and technology interaction is introduced in the Table 10.1.

Amid the transition from Industry 4.0 to Industry 5.0, emotional AI is emerging in significance as a key factor consolidating authenticity between humans and technology (Rosak-Szyrocka et al., 2025; Zywiołek et al., 2024). If the old paradigm was all about extending automation and digitalization, Industry 5.0, according to the orientations of the European Commission, is centred around a set of values including resilience, social inclusion, empathy, sustainability, and responsibility (European Commission, 2021). Institutionally, emotional AI is repositioned as not just a technology to modify the user experience, but a system that augments emotional skills and emancipation in the face of complex job situations.

It is also noteworthy that such technology has many empirically proven applications in industrial settings (Table 10.2). Ambrogio et al. (2026) discussed that EEG and heart rate variability (HRV) signals recorded from machine operators were

DOI: 10.1201/9781003754213-13

TABLE 10.1
Type of Modalities in EAI

Type	Description
Visual modality (facial expression and mimicry)	The visual domain focuses on the analysis of facial expressions, muscle movements and small variations on mimicry that are related to basic affects. Emotional Al systems often rely on methods based on the work of Ekman and Friesen (1978), who created a detailed taxonomy of facial actions that are popular in psychology and interface research. According to Picard (2000), the human face provides one of the most natural and revealing channels for emotions, therefore visual channels are determined as core in systems that aim at responding to users' emotional states.
Vocal modality (tone and rhythm of speech)	The vocal modality looks at how a person speaks—how fast, or slow, high and low pitch, pauses, rhythm and that most subtle of differences, intonation. These paralinguistic cues are related to stress. It is because emotional AI systems can recognize the change in mood by and large from the sound of speech itself, without interpreting any semantic content.
Biometric modality (physiological responses)	Biometric feature also encompasses fundamental biological data, such as heart rate variability, galvanic skin response and instantaneous structures of brain activity. As McStay (2018) points out, these signals represent natural responses to stress, cognitive load or fatigue and can thus be seen as valuable indicators of the emotional state workers find themselves in at their work places. This mode is particularly useful in situations where observation of facial expressions or displays of voice signals may be restricted or unreliable.
Behavioural modality (task-related behaviour)	The behavioural modality deals with the way a person acts whilst performing tasks; how fast they respond, how they interact with interfaces, what movements are made and the duration of activities. Calvo and D'Mello (2010) point out that these behaviours frequently reflect emotional states indirectly, such as frustration or confusion, and can hence be employed to adapt work systems and foster ergonomically appropriate interaction.
Textual modality (language use and expression)	Textual modality involves understanding the meaning and style behind writing or transcribed speech. Affective computing systems can predict the emotional tone of a message—positive, negative or neutral—and this is particularly useful in remote communication situations. Textual analysis has been among the earliest approaches to detect emotion in computational systems and continues to be relevant for testing of organizational climate (Picard, 2000).

Source: Own study based on Calvo and D'Mello (2010), Fernandez et al. (2026), George Rajesh et al. (2025).

processed by AI systems to adapt the interaction interface according to their emotional status. Under high-stress circumstances, these systems decreased operating speed, reduced the complexity of visual displays and temporarily suspended more advanced functions.

These constitute physical and cognitive ergonomics since the risk of mistakes is greatly reduced by them, increasing, therefore, occupational safety during automated procedures.

Similar approaches are under development in the area of adaptive environments. The MindSpaces project, supported by the Horizon 2020 program, experimented

TABLE 10.2
Emotional AI Applications in Workplace

Application Area	Description
Machine Operators in Industrial Environments	Emotional AI leverages biological signals like heart rate variability (HRV) and EEG signal to determine the stress and fatigue of operators. By using this information, the system self-adapts by adjusting the machine's interface, issuing breaks or reallocating tasks to lower mental load.
Adaptive Workspaces	Emotional-adaptation systems that have been created in MindSpaces project interpret environment's (e.g., lighting and temperature) and user behaviour (e.g., facial expression and motion) data to use this input dynamically for real-time reconfiguration of working conditions to increase well-being, focus and overall performance.
Human-Machine Interfaces (HMI)	Emotional AI categories about face, speech velocity as well as other emotional cues of the user and adapts its communicating approach in HMI solutions. For example, the system takes pressure off messages or adds empathy to communication.
HR Onboarding and Training Systems	Emotional AI in HR systems tailors onboarding and training experiences by detecting the emotional responses of trainees, adapting pace and content delivery according to their interest level.
Early Detection of Burnout	Emotional AI identifies indications of chronic stress and burnout at work using emotional and behavioural data to trigger support cases like restorative breaks, HR alerts or task redistribution to reduce stress.
Cobots and Human–Robot Collaboration	Emotional intelligent robots and cobots emotional AI-enhanced intelligent robots and cobots analyse an operator's emotional state, regulating the intensity of interaction, work pace or level of autonomy, improving safety, convenience and trust in human-machine interactions.

Source: Own study.

with using sensors (e.g., measuring facial micro-expressions, physiological signals and user interactions) to dynamically modulate environmental conditions (the lighting, the temperature or acoustic background) based on users' emotional profiles. The project stresses how such systems are vital for building neurodiversity-friendly workplaces—such as for employees with ADHD, high sensory sensitivity or autism spectrum disorders. It is also apparent in industry reports such as Deloitte (2023) and MindSpaces (2020), which suggest that emotional AI can foster more efficient workplace environments and improved welfare at work among cognitively diverse teams. In customer-service oriented industries, where the emotions in client-face-to-service-provider interactions are paramount, emotional AI can provide valuable assistance to both the employee and the service recipient.

The Salesforce (2025) report, which surveyed 6,500 service professionals, found that 83% thought AI had enhanced their people skills and 80% felt less stressed as a result of the technology. Such systems can monitor the emotional tone of a customer's speech, adapt communication to reflect the customer's emotional state and assist employees in coping with tough situations minimizing a risk of emotional burnout.

Another use of emotional AI is to manage emotions in distributed teams and assess organizational climate. Such systems rely on sophisticated analysis of behavioural and communication data (e.g., number of interactions, response time between interactive events) for early detection of organizational problems. For example, in the research described in the Salesforce report (2025), they report that emotion recognition during client communications facilitates early recognition of stressful interactions and helps to prepare staff for emotionally challenging encounters. Analogous mechanisms can be leveraged for communication within the team, so that managers can recognize signs of frustration, disengagement or burnout. Although evidence-based literature is limited in this specific area, the OECD (2023a) report indicates an increasing importance of technologies that allow workers' well-being to be monitored in real time as a component in managing work-related psychosocial risks.

Insights from current research on the application of AI to workplace well-being and safety provide new possibilities. AI need not only be about operational improvements; it may also help with mental health monitoring, emotional assistance and identification of psychosocial risk factors. This provides an entirely novel horizon of possibility for a human-centred vision of work. In an article by García-Madurga et al. (2024), a systematic review of the literature had shown that AI was already used for observatory of mental health, well-being programmes' supporting, psychosocial risks classifying and support and training programmes' personalizing as well: this is its practical relevance with respect to workers' quality of working life.

Interest is also growing in the embedding of affective systems and robotics in working context—in terms of cognitive ergonomics as well as physical and mental safety. Studies of human–robot collaboration (HRC) have found that such collaboration, if well designed, is able to decrease an operator's mental workload and increase comfort at work. In the most recent article by Segura et al. (2025), it was shown that when roles were well-defined and tasks carefully designed, operators engaged in universal robot systems using organizational cooperation could get better cognitive load and mental reserve results, validating the possibility of cognitive ergonomics in HRC.

The literature review indicates that monitoring workers' psychophysiological states in real time is becoming more achievable, although the authors raise questions as to the reliability, applicability and practicality of these systems (Fang et al., 2022). Despite these limitations, existing evidence demonstrates that this technology provides substantial promise for ergonomic and employee well-being interventions already.

The literature also suggests a growing interest in affective robotics, which aims to build robots that are able to detect and interpret emotions and react accordingly—not limited to healthcare or therapy but also extended in the industry and at work. A search of the literature from 2013 to 2022 (Spitale & Gunes, 2022–2022) reveals that these robots have contributed to well-being, indicating in turn that affective robotics indeed has potential as a tool for emotional support and workplace ergonomics across different work contexts.

Global technology firm Hewlett–Packard (HP) bolstered its business well-being programme Well Beyond by working with Virgin Pulse, now a Personify Health company, to provide emotional, physical and social support forward. An official

Virgin Pulse e-book, 3 Trends Driving the Future of Employee Health and Wellbeing in 2024, presents case for success of HP as client where it is indicated that the company launched a structured 90 days well-being challenge to promote employee's overall health (Virgin Pulse, 2023).

HP leveraged Virgin Pulse's AI-powered personalization engine, which utilizes machine-learning-based recommendation algorithms to personalize well-being "nudges," guided journeys and motivational messaging. AI learns over time and adjusts content, frequency and timing of well-being interactions to employees' individual requirements. It is stressed that across the entire trajectory, personalization supported by AI was infused into daily micro-experiences, ensuring that exposure to complex informational and motivational triggers did not become so burdensome as to detract from ongoing participation (Emergent Africa, 2025).

The findings of the HP-Virgin Pulse wellness programme claimed that 75% of HP employees who were eligible enrolled in the programme, which is roughly 37,000 employees worldwide. In addition, the programme "makes HP a better place to work" 97% according to participating employees. The impacts are reported to be a direct result from the introduction and availability of a corporate-specific AI Personalized Wellbeing platform that provides instant support for employee's emotional and behavioural health (Emergent Africa, 2025; Virgin Pulse, 2023).

As an example of developments for lowering workplace costs agreed at the Nissan Intelligent Factory Tochigi, towards adaptive systems which can cope with requirements and minimized physical and operational dispersion, modernization of assembly processes. The application shall illustrate how production settings can be adapted to make it easier for workers to perform their work so that this is no longer a health burden characterized by strenuous body exertions or high manual demands. Adjusting the production line to accommodate vehicles with various powertrains, such as internal combustion engines, e-POWER systems and electric-powered cars involved handling a wide variety of underfloor configurations and layout patterns. Previous assembly methods included jobs that required repetitive reaching overhead, manually aligning parts and carrying heavy pieces (in the case of electric cars, this can include the weighty battery packs). These were high physical exertion jobs that required a level of accuracy under difficult working conditions. The introduction of automation to Tochigi through a modernization project addressed some of these limitations, as it helped to create a more stabilized work environment by reducing the variability in what was required (Nissan, 2022).

A key to the changeover from overhead circuit-engaged assembly was an underfloor parts batch mounting system with a two-layer pallet construction which uses six-axis and two-axis robots for positioning between underfloor modules and vehicle bodies. The lower pallet layer serves as the universal transport structure, with the upper layer configurable to the specific model and drive-type packages. On previous jobs, they had to make manual adjustments to part support and lifter tooling to control variations in vehicle structures between different models. These manual adjustments were removed when the reconfigured system enabled robots to automatically adjust for dimensional variation, with all rear suspension assembly accuracy reviewed by vision-based measurements of vehicle geometry as well as rear suspension assembly data. The position of the vehicle on the main line is found by the vision system, and also the assembly tolerances of rear suspension positioned in sub-line pallet are calculated. The six-axis robot, after the lifting still adjusts the position of the rear shock absorbers for engaging it and turns most critical to make sure is align a pallet to vehicle body by two-axis robots. This routine has replaced time-consuming manual calibration under the vehicle and reduced one of the most physically demanding parts of the procedure (Nissan, 2022).

Further adjustments were made to reduce physical and cognitive demands of tightening operations. In traditional assembly, operators are obliged to handle heavy tools with high torques in tight environments holding them often with both hands and accurately positioning at various screwing locations. The updated line features in-line tightening with automation using long-reach tools handled by robots, backed up by a bolt reversing system that reorients the bolt before feeding it to the tightening unit. The vision-based correction process of the system also can correct for positional variation within ±0.5 mm, obviating a necessity to perform fine alignment under load or to adjust errors due to transportation and tolerance difference in parts and equipment by workers. This limits the difficulty of the task and error risk due to hand positioning.

The entire assembly of multi-link suspension modules, which is typically performed with the help of a worker using both a tightening tool and an anti-spin wrench at the same time, was redesigned as well. Nissan developed a feeding system equipped with the twin robots and utilizing the tension/resistance measurement method for control, in such a way as that one of them is fitted with nutrunner and another one with free-rotation wrench at whose axis of rotation can be found at the centre of tool. The dual-handed human operation conventional in the past is thereby simulated at this arrangement, but constraint attitudes are avoided and a stable tool setting can easily be made even in a narrow space. An automated system for clamping at multiple positions using few tools allows a small equipment size dedicated to the station and operator accessibilities according to physical diversities.

These are all solutions that help to support a production environment with fewer variations in task and more inclusiveness. Nissan says that the combination of other floor activities into one high-tech monitoring cell has helped eliminate around 15% of physically demanding operations. This decrease is attributed to the elimination of overhead work, reduced manual calibration and degradation in intensity dependence for high-load and high-accuracy tasks performed by automated equipment. Meanwhile, this did not affect production quality thanks to automatic inspection technologies and selective assembly jobs, which minimized the dependency on manual methods for tolerance compensation and allowed to minimize risks of fatigue or incorrect positions. This Nissan Intelligent Factory case is an example to showcase the possibility of adaptive technological systems, introduced in a way to assist workers with different physical characteristics and consequently reducing manual strain and providing an environment that is more stable, leading into the broadening of manufacturing role accessibility among other populations. Table 10.3 presents worker-centred design applications in workplace.

TABLE 10.3
Worker-Centred Design Applications in Workplace

Application Area	Description
Assistive industrial exoskeletons	Industrial exoskeletons, both passive and active, diminishing musculoskeletal burden by stabilizing posture and providing support to workers in repetitive or forceful operations. They contribute to decreasing fatigue and musculoskeletal risks, enabling a large number of operators with different physical abilities to access the work (Botti & Melloni, 2024).
Ergonomic human–robot collaboration (HRC)	The specifications of the HRC systems which are both physically and cognitively ergonomic. Robots are used to lower physical burden, improve posture and lessen cognitive burden of the operator while keeping a worker in control and overall supervision. Signs of benefits are seen in operator comfort, task stability and safety (Lorenzini et al., 2022).
Ergonomics-aware task scheduling (load-balancing scheduling)	Scheduling algorithms integrating the fatigue and biomechanical load of workers into the production planning. They are designed in such a way to avoid overload, to re-discharge the demanding tasks alternating high and less skilled activities and shifting workers wisely, so as to reduce the hazard derived from cumulative load (Gualtieri et al., 2020).
Multisensory or accessibility-enhanced workstations	Workplaces for workers with varied physical and sensory abilities. They incorporate multimodal cues (visual, acoustic and haptic), adaptive support and adjustments for inclusive active participation in industrial environments (Botti & Melloni, 2024).
Real-time ergonomic monitoring systems	Sensor-based platforms for real-time monitoring of posture, muscle activation (EMG), repetitive motion(s), joint angles and physical fatigue counters. These systems give workers and supervisors actionable feedback and support early interventions to reducing excessive load (Lorenzini et al., 2022).

Source: Own study.

10.2 ADAPTIVE SYSTEMS OF INCLUSION AND VARIABILITY

Within the framework of Industry 5.0, occupational safety is embedded in a more far-reaching organizational logic that focuses not only on compliance but creates a production environment that is resilient, inclusive and sustainable because of it. The European Commission underscores that Industry 5.0 shapes industry in a future with sustainable, human-centred and resilient industries where technologies enable people, rather than replacing them (Abouhawwash et al., 2024; European Commission, 2021; Kumar et al., 2024). Within this framework, the adaptive safety and workplace systems can be seen as intermediaries between technology and organizational culture and values, the design of which must draw on three core principles (resilience, inclusiveness and sustainable development) that are well established within literature (Aheleroff et al., 2022; Costa, 2024; Ghobakhloo et al., 2025; Ivanov, 2023).

Variability in work is a fundamental idea for ergonomics, socio-technical systems engineering and organizational resilience research. Variability has been described in the literature as "the natural fluctuations of performance differences, or departures, from a norm or standard" (OECD, 2023a), which include variations in work rate, posture, biomechanical load exposure level and information processing needs and sensory sensitivity/fatigue level that may fluctuate both within shifts and long term. OECD (2023c) and ILO (2023) underline that employees are not alike in terms of physical abilities, cognitive skills or sensory capabilities, and these differences introduce considerable variability into task performance which affects safety, productivity and work station design. Variety is here to stay as an intrinsic property of recent production systems and needs to be handled with adaptable solutions, particularly considering the growing complexity of Industry 5.0 scenarios.

Variability is hence not just a functional aspect of human performance but it also enters the structure for systems' adaptative safety, insofar as these systems depend on detecting and correcting such variation in situ to keep safety and human-based efficiency at adequate levels (Erik, 2017). Variability has a multifaceted source including the physical, cognitive, sensory and environmental domains. Physical performance variability is due to muscle strength, endurance and movement dynamics and fatigue (Lorenzini et al., 2022), the latter being justified by studies which address musculoskeletal load and repetitive movement activities that have shown real high variation in physiological responses to the same task. The cognitive diversity is indicative of factors such as processing speed, load, emotional regulation and vulnerability to stress discovered in studies on the modes of affective and cognitive human–computer interaction (Calvo & D'Mello, 2010; Picard, 2000). Sensory variability is a term used to describe variations in the way workers experience and respond to light, sound, colour and visual-acoustic signals as observed by studies on adaptive environmental architecture (MindSpaces Project, 2020). Environmental variabilities included changes in work conditions (e.g., noise, lighting, workplace layout or production tempo), which have already been identified as contributing the most to biomechanical risk and error probability and are principal drivers of adaptive technology use (Nissan, 2022).

Industry 5.0 adaptive systems are intended to alleviate variability and smoothen the workflow by adapting environmental conditions, portrayal of physical load and cognitive demand. Studies in human–robot collaboration have shown that robots are capable of adaptively altering trajectories, speeds and force levels thus leading to decreased postural and movement variability and more stable biomechanical load profiles among workers (Lorenzini et al., 2022). Algorithmic planning problems where load-balancing principles are entrenched have demonstrated that workers' cumulative fatigue decreases and the variability in repetitive tasks is contained by distributing more evenly, physically demanding activities among workers (Gualtieri et al., 2023). Exoskeletons used in industry, either passive or active also show a decrease in the variability of joint loading and posture with load being applied to the wearer, particularly when performing overhead work and/or elevating the arm for long periods (Hyundai Motor Group, 2019, 2020a). Studies from the automotive industry have shown that automated vision systems, six-axis robots and redesigned assembly stations reduce physical load variability and unstable dynamic movement patterns resulting in improved predictability and safety (Nissan, 2022).

From the perspective of current theories of inclusiveness, a major source of variability is by way of attentional and cognitive differences often portrayed as neurodiversity. OECD (2023b) emphasizes that environments need to be tailored for workers with an array of sensory and cognitive profiles, including those who are neurodivergent, showing a greater variance in relationship to environmental agents. Studies on humans and technology interaction have demonstrated that adaptive emotional and sensory systems—with respect to visual, acoustic or physiological modalities—may assist employees with various perceptual patterns and emotion-regulation abilities (Calvo & D'Mello, 2010; Picard, 2000). These solutions are influenced by universal design guidelines that aim for broad functional diversity, such as physical, sensory and cognitive (Breque et al., 2021). Therefore, diversity in neurocognitive profiles is not a limitation but rather an organic condition of our workforce, and its accommodations are the cornerstone of inclusive work environments that redefine advancement as we move into Industry 5.0 (Reyes et al., 2024).

The relationship between variability and organizational resilience is recognized in standards and research as fundamental to workplace safety. ISO 22316:2017 defines organizational resilience as the capability to absorb and adapt to disruptions under conditions of variable load and uncertainty, requiring work systems to remain stable despite natural fluctuations in worker behaviour and environmental dynamics (International Organization for Standardization, 2017). OECD (2023) highlights that resilience extends beyond technical infrastructure to encompass the organization's ability to anticipate and respond to diverse patterns of variability using predictive and learning-based technologies. Studies on adaptive safety management demonstrate that integrating cyber-physical systems, load-detection algorithms and biomechanical monitoring allows continuous stabilization of work and mitigation of deviations from optimal operational parameters (Bittencourt et al., 2025). Adaptive safety systems, therefore, not only compensate for variability but transform it into a mechanism that strengthens operational resilience within the Industry 5.0 paradigm.

The former comprises two dimensions: inclusiveness, which OECD and ILO (2023) put forward as the practice of creating an inclusive work environment in which all employees have physical, health-related, sensory neurocognitive, cultural or age differences are perceived. The European Commission also argues that inclusiveness refers to design of workplaces for active participation of people with diverse physical, cognitive, health and age-related conditions (European Commission, 2021). The concept is further developed in the attached documents, with descriptions of workstations that are accessible to people with various capabilities, such as multisensory and accessibility-enhanced workstations which use multimodal cues (visual, auditory and haptic) for effective task dispatching to different types of workers (Botti & Melloni, 2024). It is therefore not ovular—it is an aspect of how organizations function naturally, as supported by advice and research reports that show organizations that bring safety, diversity and well-being values in as the building blocks for designing healthy operational systems are operating with more sustainable models of operation and higher levels of performance (Deloitte, 2023).

The second is to do with organizational resilience which, as defined in ISO 22316:2017, "is the ability of an organisation to absorb and adapt in a changing environment by sustaining its achievements" (International Organization for Standardization, 2017). OECD emphasizes that the resilience of work is in terms of "the ability of technical and social systems to function effectively under maladjusted conditions" (through predictive technologies, data redundancy or learning processes)

OECD (2023b). Organizational resilience is associated with the adaptive safety models where cyber-physical systems and AI algorithms check in real-time working conditions and adjust by themselves machine speeds, tool settings, access rules to the risk zones or even to the physical (and cognitive) workload that workers are enduring while performing tasks (Bittencourt et al., 2025). Resilience becomes a technical as well as social property, applied on infrastructure and of people whose jobs could be adapted to their capabilities and levels of tiredness.

The third fundamental idea is sustainable development, a concept originally put forward by the World Commission on Environment and Development (WCED, 1987), which defined sustainable development as "development that meets the needs of the present without imposing a burden on future generations to meet their own needs." From an organizational point of view, sustainable development is a long-term ability of organizations to apply models of production that precede the economic one and take in account social and environmental aspects. This definition is broadened in EU law by the European Commission which includes employment security, health and safety, access to work and well-being as part of "social sustainability" in its document "Industry 5.0" (European Commission, 2021). In the appendices, sustainable development is connected with AI employed to support long-term worker health and capabilities: emotional-adaptation systems, predictive fatigue-prevention devices and exoskeleton techniques (Ambrogio et al., 2026; Lorenzini et al., 2022).

These three concepts can be integrated through adaptive safety and workplace systems, using technology to dynamically mould operational circumstances. The conversion of the line at Nissan Intelligent Factory in Tochigi can be cited as one instance. The modernization plan, which included the introduction of two-tier system pallets, six-axis robots and automated vision systems, resulted in a drastic reduction of high variation in physical load and overhead work inspections; minimized manual adjustment operations; built an accessible work environment for operators with various physical abilities (Nissan, 2022). By this means, adaptive systems support inclusion (through the possibility for all people with disabilities to access working life), organizational resilience (as a result of low-risk and predictable processes) and sustainable development (by averting health risks along with long-term employee support).

These conclusions are echoed by the use of H-VEX and VEX industrial exoskeletons at Hyundai Motor Group that in company materials report reducing musculoskeletal load, assisting posture and fitting users with various anthropometry (Hyundai Motor Group, 2019, 2020b). These exoskeletons provide increased workplace access by having adjustable support levels and configurations, helping people of different physical abilities (including ageing workers) and strengthening operations for challenging workstations. Ergonomics and technology-inclusive literature suggests that these are important elements for the design of environments with reduced load variability and increased accessibility (Botti & Melloni, 2024).

Ergonomic human–robot collaboration (HRC) is also one of the pillars of adaptive work systems. Studies by Lorenzini et al. (2022) reveal that HRC results in reduced physical and cognitive load accompanied by increasing task stability and safety. Also load-balancing scheduling given in Gualtieri et al. (2020) proposes production-planning algorithms that consider fatigue and biomechanical strain, safeguarding workers from cumulative overload and promoting long-term health.

A valuable addition to these systems is the use of EMG sensors to provide real-time ergonomic assessment, recording joint angles and detecting repetitions combined with an algorithm to avoid overloading and allow for immediate corrective action that reduces accidents directly and promotes organizational resilience (Lorenzini et al., 2022). Together with emotional-adaptation systems, these technologies generate workstations which react to operators' stress, fatigue and emotional conditions, addressing well-being and long-term well-being (Ambrogio et al., 2026). Table 10.4 presents adaptive safety systems in Industry 5.0 and its relevance for inclusive workplaces.

A noteworthy example of workload variability reduction and enhanced accessibility is demonstrated in the upcoming assembly line modernization under the Nissan

TABLE 10.4
Adaptive Safety Systems in Industry 5.0 and Its Relevance for Inclusive Workplaces

Type of Adaptive Safety Systems	Description	Relevance for Inclusive Workplaces
Physiological and biomechanical monitoring systems	Such systems employ wearable sensors, cameras and machine-learning algorithms to analyse fatigue, musculoskeletal load, postural adaptations and physiological data (e.g., heart rate variability and muscle activation). They evaluate the status of the operator dynamically and report feedback to process control systems.	They also make it possible to continuously adjust the intensity and speed of work according to the operator's ability level, resulting in less overload and risk of injury. They aid those of mixed physical capabilities, including older employees or workers recovering from injuries to make the workplace more accessible and to remove obstacles to participation.
Industrial exoskeletons (active and passive)	The human body is therefore supported by these exoskeletons, which mechanically relieve the above-mentioned joints, stabilize posture and perform part of the necessary work. Passive structures rely on mechanical stabilization, while the active ones apply actuation and powered support for automatic response to user movement and load.	They minimize fatigue, restrict joint-torque exposure and stabilize imposed posture in confined spaces. In the Industry 5.0, it ensures seniors and people with less physical strength can work longer, promoting diversified availability of work opportunities.
Human–robot collaboration (HRC)	The tasks in HRC are divided between the robot which takes care of physical, repetitive or dangerous activities and the worker who keeps supervising and decision-making responsibilities. Robots have safety sensors and a compliant control to adjust the path, velocity or force when they sense human existing.	They allow work for persons of different physical strength, decrease forced movements and even out the workflow. They improve predictability and access to the work environment, reduce cognitive load by helping to complete complex tasks.

(Continued)

TABLE 10.4 (CONTINUED)

Type of Adaptive Safety Systems	Description	Relevance for Inclusive Workplaces
Adaptive multisensory workstations	These workstations make use of systems that are able to adapt the height, reach and visual, auditory and haptic stimuli. With the aid of motion sensors, cameras and user interfaces, the system "learns" operator preferences and capabilities and, on its own, revises information layout and operational settings.	They also enhance spatial orientation, decrease the risk of mistakes and assist employees with different sensory requirements like those who are visually impaired or hard of hearing, as well as neurodiverse candidates. It promotes equitable participation in the industry.
Adaptive task-planning and job-rotation models	Scheduling algorithms consider the physical and psychosocial workload, fatigue, work load capacity and individual constraints to optimize the organization of work shifts. They design tasks such as task cycles with different difficulties or adjust the order of operation automatically on a production line.	They also mitigate overexertion from chronic exposure to engaging tasks and may allow the participation of workers with various health profiles, ages or caring responsibilities. They want workplaces that are more flexible and fair.
Digital-twin-based safety architectures	Workstation and production line digital twins are populated with sensor-based, real-time data on movement, load, machine parameters and worker behaviours. Simulators foresee crashes, breakdowns or overload and reset automatically the safety conditions.	They enable experimental risk-free performance of organizational alternatives and the assessment of access and safety in a places or workplace for people with different abilities. They enable more predictable, stable and user-driven worlds of work.
AI-driven adaptive safety protocols	Risk-analysis algorithm, fault prediction and hazard classification algorithms: these exploit environmental, machine and operator data to dynamically adapt safety rules (safe-guard distances, alert density and robot speed limits). Such systems adapt immediately to the new conditions.	They enhance the accuracy and promptness of interventions, minimize incidents or false positives and address alerts for different user types to make it easier on workers with sensory deficits/cognitive impairments.

Source: Own study.

Intelligent Factory. The utilization of the two-level palette system has in principle made it possible to remove all work overhead, which is from biomechanical point of view one of the largest sources for the increased risk for musculo-skeletal fatigue. Replacement of force-raising tasks, inclusion of fully automatic batch mounting, and using six-axis robots decreased posture variation and constrained dynamic changes in the physical load magnitude and sequence pattern (thereby enhancing predictability) as well as standardized the operation procedure (with increased safety). The technical literature specifies that these solutions not only standardize the working parameters but also allow the workstations to be accessible, regardless of operator strength, age and motor potential, in accordance with the principles for designing work environments which meet functional diversity (Nissan Motor Corporation, 2022).

10.3 HEALTH AND SAFETY WITHIN ESG AND HUMAN CAPITAL POLICIES

In the context of Industry 5.0, occupational health and safety no longer sit discretely as legalistic compliance activities but are now embedded in corporate responsibility materiality assessments, human capital strategies and ESG governance models contributing to service organization value drivers (Kvasničková Stanislavská et al., 2023; Xue et al., 2025; Zywiołek et al., 2024). Health and Safety of workers they are increasingly regarded as fundamental drivers of long-term organizational resilience, stakeholder trust and sustainable business performance (Deloitte, 2023; OECD, 2023a). Independent ESG rating organizations such as MSCI and Sustainalytics characterize OHS, safety, well-being and labour conditions as material social risks affecting the enterprise value, operational continuity and long-term resilience of the organization (MSCI, 2023; Sustainalytics, 2024). From this perspective, employees physical and mental state of health, accessibility to workplaces or the ethical use of new technologies can be regarded as general elements of social value generated by companies (Zywiołek et al., 2024).

The ESG paradigm extends the classic notion of occupational safety to non-financial reporting obligations by considering absenteeism, turnover due to health reasons, psychosocial risks and quality of environmental conditions (or availability of inclusive ergonomic work stations) as expressions of social responsibility (European Commission, 2021; García-Madurga et al., 2024). Through the implementation of the Corporate Sustainability Reporting Directive (CSRD), companies must report on their impact on employee health and well-being, which raises OHS from compliance in HR when revolved by Health and Safety at Work Act to a strategic aspect of human-capital governance (European Commission, 2021). As a result, occupational safety becomes an observable and reportable societal outcome that is linked to firm legitimacy and governance performance.

According to the GRI 403 Occupational Health and Safety standard, worker health, safety, psychological risks such as bullying or harassment and safe access to ergonomically designed work environments are considered as material core topics for sustainability reporting underlining that OHS performance is an integral part of a corporate social responsibility (Global Reporting Initiative, 2021). ESG reporting frameworks are now demanding more from organizations in reporting on safety performance indicators like LTIFR, TRIR, absenteeism related to work perimeters, turnover rates attributed to health issues and validated well-being measures, which all are increasingly seen as key proxies for long-term social sustainability and human-capital value creation (Deloitte, 2023).

In human-capital management literature, there is a growing understanding of the need for incorporating safety and well-being into talent strategies, competencies in development frameworks and long-term workforce sustainability plans (Deloitte, 2023; Salesforce, 2025). According to the "Global Human Capital Trends 2023" report, mental health and emotional well-being as well as having access to enabling working conditions are underlying assets which contribute towards impairing the long-run value of human capital (Deloitte, 2023). Thus, emerging well-being ecosystems, burnout-prevention programmes and personalized health interventions constitute elements of HR agendas with digitally enabled/managerial policy coordination.

According to systematic reviews, AI-based well-being systems might contribute to the early detection of psychosocial risks as well as intervention personalization and real-time emotional load monitoring (García-Madurga et al., 2024). These technologies enable a fast stress pattern detection, the classification of psychosocial threats and targeted support, which appears consistent with the "S" aspect of ESG from the perspective that it is a duty of care to protect physical and mental health ((OECD, 2023a). These systems act as a preventive tool for HR and OHS professionals, informing organizations of potential hazards. Organizations have a legal obligation to prevent psychosocial risks and manage work-related stress and to implement evidence-based mental health interventions (World Health Organization & International Labour Organization, 2022), rendering emotional well-being an obligatory aspect of the social component of ESG requirements.

A need for ethical governance of technology is also highlighted in OECD reports as well as in research on responsible AI, where there is an emphasis on the importance to protect privacy, keep algorithms transparent and prevent too much surveillance (McStay, 2018; Picard, 2000). On the governance dimension of ESG, this involves principles of proportionality, transparency and data minimization with respect to management of biometric and psychophysiological data (OECD, 2023c). Consistent with "responsible AI" principles, workplace technologies ought to strengthen employees' agency and autonomy rather than diminish them.

Studies on human–robot collaboration (HRC) stress that the well-being can be enhanced through technological systems when considering this within human-centred ergonomics and cognitive-design frameworks (Segura et al., 2025). When embedded in ESG reporting, these systems serve to both re-enforce efficiency and support the workers, thus bolstering long-term workforce capacity. From human-capital perspective, this change also suggests that the organizations need to manage not only regarding productivity but being how to maintain physical and emotional health as well as employability of employees in a long cycle of their careers (Hermawati et al., 2025).

Industry reports place the focus on a personalized well-being ecosystem in which AI powers custom health-support pathways and micro-interventions around workers' real-time states. These include the HP Well-Beyond programme and, AI-enhanced well-being platforms such as that rolled out by Virgin Pulse, which drive participation and embed preventive organizational cultures (Emergent Africa, 2025; Virgin Pulse, 2023). These activities are directly related to the S and G aspects of ESG, as they increase employee involvement, transparency and corporate caring.

The notion of "social sustainability" in Industry 5.0, as identified by the European Commission, extends occupational safety to long-term functional capacity and emotional fitness in a bias-resistant use of work-enabling technologies over the life course (European Commission, 2021). The literature on Industry 5.0 further highlights that "organisations establishing healthy, well and safety into their ESG frameworks affect not only the organization's ESG performance but also its stability and resilience" (Aheleroff et al., 2022; Costa, 2024).

From a governmental point of view, the regulatory standards for example ISO 45001 (occupational health and safety management) and ISO 22316 (organizational resilience) are important basis to show conformity and other ways to combine risk management systems with ESG reporting activities (International Organization for

Standardization, 2017). When included in ESG frameworks, these standards serve as proof of preventive measures and facilitate external auditing and organizational maturity (Bittencourt et al., 2025).

The integration of health, safety, and well-being into ESG and human-capital strategies repositions these areas as strategic levers for responsible, resilient and socially sustainable Industry 5.0 organizations. Technologies such as AI, robotics, psychophysiological sensing and WellTech are not mechanisms of substitution but of augmentation—ethical, inclusive and human-centred workplaces (Kumar et al., 2024; Xue et al., 2025). The literature unequivocally shows that in Industry 5.0, occupational safety is no longer reactive regulatory dimension but ingrained in the corporate value creation process, forming a basis for organizations to operate within its licence to work, being sustainable and resilient over time (Abouhawwash et al., 2024; Hermawati et al., 2025; OECD, 2023b; Zywiołek et al., 2024). Table 10.5 presents relationships among ESG, occupational safety and human-capital policies in the era of Industry 5.0

TABLE 10.5
Relationships among ESG, Occupational Safety, and Human-Capital Policies in the Era of Industry 5.0

ESG/Human-Capital Area	Role in ESG	Tools and Examples
Environmental (E)	The materiality of workplace environmental conditions, which impact in a direct way on the physical and mental health of workers; stage of environmental exposure as part of responsible manufacturing.	Air monitoring, noise/vibration dampening systems, microclimate analysis, incorporating environmental exposure in OHS risk assessment.
Social (S): Health, Safety & Well-Being	Central domain in which health, safety, well-being, inclusivity and good work are situated; closely associated with human sustainability.	Well-being portals (e.g., Virgin Pulse), psychosocial risk tracking, ergonomically inclusive workstations, age-inclusive planning for design, health promotion programmes.
Governance (G): Responsible Technology & Compliance	Ethical, transparent and compliant handling of OHS data, AI, robotics and organizational risks: a corporate responsibility issue.	ISO 45001, ISO 22316, GDPR, OECD AI Principles, algorithmic transparency, ethical audits.
Human Capital Policies	HR strategies that create enduring workforce sustainability, functional capability, health, emotional resilience and retention.	Personal well-being journeys AI-based workload monitoring burnout prevention learning safe use of technology talent retention programmes.
Industry 5.0 Synergy	Utilizing both ESG and human capital in combination with human-centric technology to create responsible, resilient businesses.	Cyber-physical safety systems, digital twins for human health and well-being, ergonomic interactive robots and adaptable workplaces.

Source: Own study.

CASE STUDY SIEMENS – HEALTH, SAFETY AND WELL-BEING INTEGRATION INTO ESG AND INDUSTRY 5.0

One of the good examples of how a global company can put ESG and human capital programme at an integrated system along with employee health and safety is Siemens. The current Siemens Sustainability Report 2024 explicitly says that occupational health and safety management steers towards resilience and adaptability as well as employee welfare and underpins that "resilience and welfare are built into occupational health and safety management." This represents a shift from the instrumentalist and reactionary conception of OHS derived from risk and regulation to recognizing health, safety and well-being as critical forms of intangible capital. Siemens specifically mentions "health, safety and resilience" as "intangible assets for our company," embedded in the strategic focus on "empowered people," connecting OHS KPIs to action fields of their DEGREE sustainability framework (mainly in the "Employability" field: 30% reduction target of global LTIFR by 2025 compared to 2020).

The Siemens OHS management system is part of a wider ESG governance structure with the responsibility for this business function being assigned to the EHS (Environmental Protection, Health Management, Safety) organization that exists as a site-based or regional company-based component reporting directly to these managements. The EHS officers coordinate the activities of health and safety professionals, and every operating unit has to have a management system in place that can be certified according to ISO 45001; the effectiveness thereof is being reviewed annually by means of internal audits and, if market demands, externally (certification). In this way, OHS is incorporated into governance: it is connected to the internal monitoring of the group, to enterprise risk management and the "Business Conduct Guidelines," and is also transferred across supply chains through a——Code of Conduct for Suppliers and Third-Party Intermediaries, which imposes standards on suppliers with regards to OHS and makes them responsible for the well-being of their employees.

Siemens has implemented the "Healthy and Safe @ Siemens" (HS@S) programme worldwide. This programme is grounded in five values: caring for your own well-being and that of your colleagues; make sure employees can have their say about safety concerns, integrate different perspectives, leverage knowledge on a better way to work and prepare for shifts with change. HS@S is based on a "reality check" instrument to diagnose staff perceptions in the dimensions of work climate, leadership, learning and processes and resources and lays out measures for "resilient health & safety management" following this with an evaluation of perceived effects 2 years after taking action. Like any good programme, Siemens has gradually been broadening the complementary health services it provides its employees: in 2023, they have made the benefit available for 99% of their workforce to in-house medical care which includes prevention and early detection of health problems (examining boards, screenings and vaccination) with quick access to specialists; the telemedicine structure is also being strengthened. In parallel, global initiatives as "Be well + Work Well" and "Check in for Mental Health" have been implemented, where psychological well-being is fostered already since operational risk management focused on psychosocial risks within the hazard identification, risk assessment, and risk control (HIRARC) process (Hazards Identification, Risk Assessment and Risk Control) all year long, along with access to Employee Assistance Program (EAP), offering individual support on psychosocial issues and personal needs.

It is worth to mention that Siemens also applies conventional accident rate indicators and numerous factors related to ESG policies. One of the targets within the DEGREE sustainability model is to reduce the global LTIFR by 30% by 2025 (from a base of 0.31 in 2020) and for employees, it improved by 19% compared to baseline with an overall value of 0.25 between employees and temporary workers (and only 0.22 for direct employees). Occupational diseases are also mentioned (in 2024, 74 were reported in which musculoskeletal suffered the most) as well as fatal accidents, and each is seen as a continuation of engineering system upgrade. Siemens combines psychosocial risk assessment with its overall human capital governance system/annual measurement of workplace well-being factors by means of their global Siemens Global Engagement Survey, including a structured follow-up process at the team level intended to identify risks, opportunities and corrective actions which are documented for reference.

REFERENCES

Abouhawwash, M., Rosak-Szyrocka, J., & Gupta, S. K. (2024). *Aspects of Quality Management in Value Creating in the Industry 5.0 Way*. CRC Press.

Aheleroff, S., Huang, H., Xu, X., & Zhong, R. Y. (2022). Toward sustainability and resilience with Industry 4.0 and Industry 5.0. *Frontiers in Manufacturing Technology*, *2*, 951643. https://doi.org/10.3389/fmtec.2022.951643

Ambrogio, G., Longo, F., Mirabelli, G., & Padovano, A. (2026). Inside-out: An exploratory study of the emotional profile of industrial operators for human-centric factories and empathic AI assistants. In *Advances in Production Management Systems. Cyber-Physical-Human Production Systems: Human-AI Collaboration and Beyond* (pp. 194–208). Springer.

Bittencourt, V., Kolla, S., Saakes, D., & Thiede, S. (2025). Towards adaptive human-centered workplaces – framework & building blocks. *Procedia CIRP*, *134*, 813–818.

Botti, L., & Melloni, R. (2024). Occupational exoskeletons: Understanding the impact on workers and suggesting guidelines for practitioners and future research needs. *Applied Sciences*, *14*(1), 84. https://www.mdpi.com/2076-3417/14/1/84

Breque, M., De Nul, L., & Petridis, A. (2021). *Industry 5.0 – Towards a Sustainable, Human-centric and Resilient European Industry*. European Commission, Directorate-General for Research and Innovation. https://op.europa.eu/en/publication-detail/-/publication/468a892a-5097-11eb-b59f-01aa75ed71a1

Calvo, R. A., & D'Mello, S. (2010). Affect detection: An interdisciplinary review of models, methods, and their applications. *IEEE Transactions on Affective Computing*, *1*(1), 18–37.

Costa, E. (2024). Industry 5.0 and SDG 9: A symbiotic dance towards sustainable transformation. *Sustainable Earth Reviews*, *7*(1), 4. https://doi.org/10.1186/s42055-024-00073-y

Deloitte (2023) *Human Capital Trends – New Fundamentals for a Boundaryless World.* https://www.deloitte.com/ca/en/services/consulting/research/2023-global-human-capital-trends-report.html

Ekman, P., & Friesen, W. V. (1978). *PsycTESTS Dataset*: American Psychological Association (APA).

Emergent Africa (2025), *Case Studies of Successful AI Implementation in Employee Wellness.* https://www.linkedin.com/pulse/case-studies-successful-ai-implementation-employee-wellness-8svgf

Erik, H. (2017). *FRAM: The Functional Resonance Analysis Method*: CRC Press.

European Commission. (2021). *Industry 5.0 – Towards a Sustainable, Human-Centric and Resilient European Industry* (Publication No. KIBD20021ENN). Publications Office of the European Union. https://op.europa.eu/s/qq0Y

Fang, W., Wu, D., Love, P. E., Ding, L., & Luo, H. (2022). Physiological computing for occupational health and safety in construction: Review, challenges and implications for future research. *Advanced Engineering Informatics*, *54*, 101729.

Fernandez, G., Assis Dornelles, J. de, & Ayala, N. F. (2026). How Industry 4.0 technologies are evolving to Industry 5.0 - A systematic literature review. In *Advances in Production Management Systems. Cyber-Physical-Human Production Systems: Human-AI Collaboration and Beyond* (pp. 48–61). Springer.

García-Madurga, M.-Á., Gil-Lacruz, A.-I., Saz-Gil, I., & Gil-Lacruz, M. (2024). The role of artificial intelligence in improving workplace well-being: A systematic review. *Businesses*, *4*(3), 389–410.

George Rajesh, S., Vipin Madangarli, S., Santosh Pisharady, G., & Subrahmanyam, R. (2025). Enhancement of virtual assistants through multimodal ai for emotion recognition. *IEEE Access*, *13*, 102159–102179.

Ghobakhloo, M., Fathi, M., Okwir, S., Al-Emran, M., & Ivanov, D. (2025). Adaptive social manufacturing: A human-centric, resilient, and sustainable framework for advancing Industry 5.0. *International Journal of Production Research*, 1–34.

Global Reporting Initiative. (2021). *GRI 403: Occupational Health and Safety 2018* (updated 2021). https://www.globalreporting.org

Gualtieri, L., Palomba, I., Merati, F. A., Rauch, E., & Vidoni, R. (2020). Design of human-centered collaborative assembly workstations for the improvement of operators' physical ergonomics and production efficiency: A case study. *Sustainability*, *12*(9), 3606.

Gualtieri, L., Rauch, E., & Vidoni, R. (2023). Human-robot activity allocation algorithm for the redesign of manual assembly systems into human-robot collaborative assembly. *International Journal of Computer Integrated Manufacturing*, *36*(2), 308–333.

Hermawati, S., Correa, R., Mohan, M., Lawson, G., & Houghton, R. (2025). Defining human-centricity in Industry 5.0 and assessing the readiness of ergonomics/human factors communities in UK. *Ergonomics*, *68*(6), 857–876.

Hyundai Motor Group. (2019, September 3). *Hyundai Motor Group Develops Wearable Vest EXoskeleton to Alleviate Burden in Overhead Work.* Press Release. https://www.hyundai.news/uk/articles/press-releases/hyundai-develops-wearable-vest-exoskeleton-for-overhead-work.html

Hyundai Motor Group. (2020a). *Wearable Robotics*. Hyundai Motor Europe. https://www.hyundai.com/eu/en/mobility-and-innovation/technology-and-innovation/robotics/wearable-robotics.html

Hyundai Motor Group. (2020b, March 30). *Hyundai Motor Group's Latest Exoskeleton Wearable Robot Wins Red Dot Design Award for Innovative Product.* Automotive World. https://www.automotiveworld.com/news-releases/hyundai-motor-groups-latest-exoskeleton-wearable-robot-wins-red-dot-design-award-for-innovative-product/

ILO. (2023). *Working Time and Work-Life Balance Around the World.* https://www.ilo.org/publications/working-time-and-work-life-balance-around-world

International Organization for Standardization. (2017). *ISO 22316:2017 Security and Resilience — Organizational Resilience — Principles and Attributes*. ISO. https://www.iso.org/standard/50053.html

Ivanov, D. (2023). The Industry 5.0 framework: viability-based integration of the resilience, sustainability, and human-centricity perspectives. *International Journal of Production Research, 61*(5), 1683–1695.

Kumar, N., Pal, S. K., Agarwal, P., Rosak-Szyrocka, J., & Jain, V. (2024). *Human-Machine Collaboration and Emotional Intelligence in Industry 5.0*: IGI Global.

Kvasničková Stanislavská, L., Pilař, L., Vogli, X., Hlavsa, T., Kuralová, K., Feenstra, A., et al. (2023). Global analysis of Twitter communication in corporate social responsibility area: Sustainability, climate change, and waste management. *PeerJ Computer Science, 9*, e1390.

Lorenzini, M., Lagomarsino, M., Fortini, L., Gholami, S., & Ajoudani, A. (2022). Ergonomic human-robot collaboration in industry: A review. *Frontiers in Robotics and AI, 9*, 813907.

McStay, A. (2018). *Emotional AI: The Rise of Empathic Media*. SAGE Publications Ltd.

MindSpaces Project. (2020). *Emotion-Adaptive Workspaces*. Horizon 2020 Innovation Action. https://mindspaces.eu/

MSCI. (2023). *MSCI ESG Ratings Methodology*. MSCI Research. https://www.msci.com/esg-ratings

Nissan. (2022). *Nissan Technical Review 2022, No. 88*. Nissan. https://www.nissan-global.com/EN/TECHNICALREVIEW/PDF/NISSAN_TECHINICAL_REVIEW_88_En_ALL.pdf

Nissan Motor Corporation. (2022). *Nissan Technical Review, No. 88: Nissan Intelligent Factory Starting from ARIYA*. Nissan Motor Corporation. https://www.nissan-global.com

OECD. (2023a). *Using AI in the Workplace: Opportunities, Risks, and Policy Responses*. Organisation for Economic Co-operation and Development. https://www.oecd.org/content/dam/oecd/en.publications/reports/2024/03/using-ai-in-the-workplace_02d6890a/73d417f9-en.pdf

OECD. (2023b). *Artificial Intelligence in Work, Innovation, Productivity and Skills*. https://www.oecd.org/en/about/programmes/ai-in-work-innovation-productivity-and-skills.html

OECD. (2023c). *Using AI in the Workplace: Opportunities, Risks and Policy Responses*. https://www.oecd.org/en/publications/using-artificial-intelligence-in-the-workplace_840a2d9f-en.html

Picard, R. W. (2000). *Affective Computing*: MIT Press.

Reyes, A., Torres, S., Robinson, T. L., Rentería, M. A., & Clark, A. L. (2024). Integrating diversity, culture, and equity into the neuropsychological evaluation. *Neurologic Clinics, 42*(4), 959–979.

Rosak-Szyrocka, J., Gupta, S. K., Shahbaz, M., & Launer, M. (2025). *Enhancing Smart Universities with Emotional Intelligence: Competences for Digital Age Learning*: Taylor & Francis.

Salesforce. (2025). *State of Service* (7th ed.). Salesforce Research. https://www.salesforce.com/resources/research/state-of-service

Segura, P., Lobato-Calleros, O., Soria-Arguello, I., & Hernández-Martínez, E. G. (2025). Work roles in human–robot collaborative systems: Effects on cognitive ergonomics for the manufacturing industry. *Applied Sciences, 15*(2), 744.

Spitale, M., & Gunes, H. (2022–2022). Affective robotics for wellbeing: A scoping review. In *2022 10th International Conference on Affective Computing and Intelligent Interaction Workshops and Demos (ACIIW)* (pp. 1–8).

Sustainalytics. (2024). *ESG Risk Ratings Methodology*. Morningstar Sustainalytics. https://www.sustainalytics.com

Virgin Pulse (2023), *3 Trends Driving the Future of Employee Health and Wellbeing in 2024 – Innovative HR Strategies for 2024.*

World Commission on Environment and Development. (1987). *Our Common Future*. Oxford University Press. https://sustainabledevelopment.un.org/content/documents/5987our-common-future.pdf

World Health Organization, & International Labour Organization. (2022). *Mental Health at Work: Policy Brief*. WHO & ILO. https://www.who.int/publications/i/item/9789240053052

Xue, Y., Zhang, C., Mentel, G., Mohammed, K. S., Rosak-Szyrocka, J., & Zhao, X. (2025). Impact of ESG rating divergence on corporate resilience: insights from China's A-share market. *Environment, Development and Sustainability*, 1–26. https://doi.org/10.1007/s10668-025-06799-9

Zywiołek, J., Rosak-Szyrocka, J., Nayyar, A., & Naved, M. (2024). *Modern Technologies and Tools Supporting the Development of Industry 5.0*. CRC Press.

Index

Pages in ***bold-italics*** refer to boxes and pages in **bold** refer to tables.

Q

R

S

T

U

V

W

X

For Product Safety Concerns and Information please contact our EU representative GPSR@taylorandfrancis.com
Taylor & Francis Verlag GmbH, Kaufingerstraße 24, 80331 München, Germany

www.ingramcontent.com/pod-product-compliance
Ingram Content Group UK Ltd.
Pitfield, Milton Keynes, MK11 3LW, UK
UKHW020203200926
475702UK00004B/596

* 9 7 8 1 0 4 1 2 6 9 5 2 6 *